BEST FACE
FORWARD

BEST FACE FORWARD

*Why Companies Must
Improve Their Service
Interfaces with Customers*

Jeffrey F. Rayport
Bernard J. Jaworski

HARVARD BUSINESS SCHOOL PRESS
BOSTON, MASSACHUSETTS

Library of Congress Cataloging-in-Publication Data

Rayport, Jeffrey F.
    Best face forward: why companies must improve their service interfaces with customers / Jeffrey F. Rayport, Bernard J. Jaworski.
        p.   cm.
    Includes bibliographical references.
    ISBN 0-87584-867-2
    1. Customer services—Management. 2. Service industries—Technological innovations—Management. 3. Competition. I. Jaworski, Bernard J. II. Title.
    HF5415.5.R393 2004
    658.8'12—dc22

                                                                2004002104

CONTENTS

Acknowledgments   vii

Introduction   xiii

1. Interfaces As the Next Frontier of
   Competitive Advantage   1

2. The Interface Imperative   27

3. The Front-Office Revolution   55

4. What People Do Best   87

5. What Machines Do Best   113

6. Putting the Amalgam of People and
   Machines to Work   145

7. Managing Interface Systems   173

8. The Interface Audit   207

Notes   229
Glossary of Key Terms   249
Index   253
About the Authors   263

The Limited; and John J. Sviokla, formerly of HBS and now vice chairman at DiamondCluster. We thank our former case writers at HBS—Thomas C. Gerace (a cofounder of BeFree, Inc. and Value-Click, Inc. and now senior vice president at National Leisure Group), Carrie Johnson (now an e-retail analyst at Forrester Research), and Dickson Louie (now a consultant at the *San Francisco Chronicle*)—who created many of the field-based materials from which our observations are drawn.

And we are grateful to the many entrepreneurs who opened their doors to us for purposes of allowing us to develop case studies for our classrooms in recent years. First and foremost, we owe a special debt of gratitude to Darlene Daggett, president of QVC's North American operations, for allowing us to follow, over a period of several years, the world's leading TV home shopping business, and we thank her colleagues who supported our efforts, including Doug Briggs, Bill Costello, John Hunter, Doug Rose, and Marnie McGinnis. We also wish to express our appreciation for the support of Warren Adams of PlanetAll, which was acquired by Amazon.com; Jeff Bezos at Amazon.com; Jim Griffin, of Geffen Records and now Cherry Lane Digital; Andrew Heyward at CBS News; Larry Kramer at CBS MarketWatch; Kevin Newman at First Direct in the United Kingdom; Martin Nisenholtz at New York Times Digital; Robert Rodin of Marshall Industries and now eConnections; Carl Rosendorf of BarnesandNoble.com and now CEO of SmartBargains.com; Paul Tarvin of Frontgate Catalog; Jeff Taylor of Monster Interactive; and Royce Yudkoff and Andrew Banks of ABRY Partners. We are also grateful to Tim DeMello, of Streamline.com; Peter Granoff and Robert Olson, the "Cork Dorks" of Virtual Vineyards; Bob Jamieson and Jack Rovner, of RCA Records; Mike Leavitt of Weather Services Corp.; Stuart Spiegel, of iQVC; and Jeremy Verba of E! Online.

We owe a special note of gratitude to our heroic editor, Kirsten D. Sandberg, at Harvard Business School Press, who stuck by this project through thick and thin and who blessed our work with her ready humor, her sharp wit, her encyclopedic knowledge of the lit-

visory practice; and Mark T. Thomas, head of Monitor's merchant banking group. Other senior Monitor partners to whom we are grateful for intellectual contributions are Jennifer Barron, Jonathan Goodman, Ralph Judah, David Levy, Robert B. Lurie, Bansi Nagji, George Norsig, and our resident game-theory guru, Bhaskar Chakravorti. Without all of their practical wisdom and intellectual insights, tempered by years of creating opportunities and solving problems in the corporate world, we could not have framed our ideas as effectively for practical impact.

Among our colleagues at Marketspace, we are grateful to Katarina Gizzi for her enthusiastic research support in pursuit of many of the book's more complex arguments; to Kristia DeRoche, for help with fact checking; to Kathy Ivanciw, for her sanity checks on the progress of the manuscript and our lives; and to JoAnn Kienzle, for her editorial comments and contributions regarding continuity. Our in-house historian, acting legal counsel, and resident curmudgeon, Steve Szaraz, supported our efforts with everything from nineteenth-century literary references to vetting of legal contracts. Our network czar, Ian Findlay, kept us honest with respect to technology issues. Our Santa Monica–based "L.A. connection," Allison Reese, coordinated schedules and helped keep this project on track. Indeed, the entire Marketspace organization played a role in creating a context in which we might ask challenging questions and receive intellectually astute feedback in return, full of rich insights, real-world anecdotes, and innovative ideas, all based on a community of superbly talented people steeped in media, technology, and networks.

We are also grateful to colleagues from academic life who have played important roles in discussing these ideas with us, including Tom Eisenmann, James L. Heskett, V. Kasturi Rangan, Al Silk, and Gerald Zaltman of HBS; Ajay Kohli of Emory University; Gary W. Loveman, formerly of HBS and now CEO of Harrah's Entertainment; Deborah MacInnis (and several of her Ph.D. students) at the Marshall School of Business at USC; Leonard D. Schlesinger, formerly of HBS and now chief operating officer at

erature, and her astute editorial judgment. Also of HBS Publishing we thank our senior production editor Susan Francis and our copyeditor Michelle Cain.

And we thank our literary agent Rafe Sagalyn, for his expert guidance and unflagging commitment throughout the process of creating this book.

Finally, we thank our families for their ever-faithful and unstinting support of our work in all regards, from our respective homes in Boston and Los Angeles. Any and all infelicities in this work are, of course, attributable to us and us alone.

set of ideas focused on interactive marketing, which brought our thinking on interface systems together with how marketing campaigns work. Leading both efforts was our good friend Toby Thomas, a senior partner at Monitor Group and the head of applications development for Market2Customer, a business unit of Monitor Group specializing in customer understanding and insight. In applying ideas to business opportunities and challenges, our advisory colleague Colin Gounden has led the way in client settings. And to bring these ideas to life in executive education programs, our practice leaders Karin Stawarky and Michelle Toth in our business unit, Monitor Executive Development, inspired us with engaging ways to bring our thoughts to life with practitioner groups. Without the contributions of these extraordinary friends and colleagues, the quality of this book would have been greatly diminished.

This book was written at the intersection of academic lives that we both led previously and the corporate lives we now share as partners in Monitor Group and leaders of our business at Marketspace. The ability to pursue a project of this kind was enabled by the indefatigable loyalty, unflagging support, and expansive spirit of Monitor's founder and chairman, Mark B. Fuller, a one-time member of the HBS faculty and subsequently an entrepreneur who created one of the world's premier strategy firms. Mark is the leader who, with his cofounders, is driving Monitor Group in its successful forays beyond the strategic advisory business into private equity, venture capital, technology innovation, software development, and executive education, to name just a few of the firm's burgeoning growth trajectories. Mark's belief in our ideas did much to make this book happen. Our special thanks goes as well to William M. McClements, who supported us first as Monitor's chief operating officer and, more recently, as chief executive officer of our organization. In playing such key roles in enabling our work, Mark and Bill brought support from other key leaders at our firm: Joseph B. Fuller, head of Monitor's global advisory practice; Stephen M. Jennings, head of Monitor's North American ad-

ACKNOWLEDGMENTS

In developing the ideas for this book, we have been blessed by a talented array of colleagues within our organization, Marketspace LLC, and our parent company, Monitor Group, the strategic advisory and merchant banking firm based in Cambridge, Massachusetts. In addition, we have benefited tremendously from our colleagues in faculty posts at both Harvard Business School (HBS) and the Marshall School of Business at the University of Southern California (USC), where we served on the faculties, and colleagues elsewhere in both business and academe. We owe much inspiration and direction to the extraordinary students we had the privilege to teach at Harvard and USC in the M.B.A. programs at both those schools; their intuitive sense for technology and its changing roles in business brought energy, momentum, and inspiration to our thinking.

We are indebted, in particular, to our inspirational colleague Eleanor J. Kyung for her brilliant ideas, creative contribution, passion for technology, and tireless support of our research and writing process. Alongside Ellie, we owe a debt of gratitude to the contributions of two "applications" teams of strategic advisory professionals within our organization. The first team, whose members included George Eliades, Dorsey McGlone, Silvia Springolo, and Ellie, developed ideas around what we call *interface advantage*, a methodology that has guided us in optimizing interfaces for corporations around the world. The second team, made up of Steven Libenson, Melissa Pennings, and Aboud Yaqub, developed a related

INTRODUCTION

The world of services is undergoing a revolution. All around us, companies are radically reconfiguring the ways they interact with customers. That, in turn, is changing irrevocably how people employed in all manner of service positions—which is most of the work force—relate to customers and their jobs. In airports, computerized kiosks dispense boarding passes and automated scanners read them at the gate. In concourses, fully automated store-in-a-box vending machines bearing retail brands sell books for WHSmith and office supplies for Staples. In drugstores, Kodak-branded kiosks with brightly colored touch-screens download digital images from cameras and mobile phones and print them on demand. In large-format retailers such as The Home Depot, self-checkout stations tally up shoppers' purchases in nearly a thousand of the chain's U.S. stores. In movie theaters, Fandango kiosks dispense prepaid movie tickets and sell new ones. Call these machines the offspring of the automated teller machine (ATM), but they bear little resemblance to their cash-dispensing forebears that originated several decades ago. You might argue that what's happened in the interim—along with the underlying evolution in enabling technologies—is the Web, a mass-market training ground for consumers in dealing with the symbolic logic of point-and-click icons, pull-down menus, hyperlinked content, and electronic contexts for accessing services and transactions. Mass-market consumers of all ages and walks of life have embraced these machines—and adaptation of these consumers to new ways of interacting with the world has taken place with remarkable velocity.

The result is a revolution in the service sector that's hard to understate: The front office has succumbed to automation in much the way manufacturing did over a century ago. The shift from human to machine labor in mass-market services resembles earlier industrial revolutions, when steam-driven turbines replaced living muscle in the early nineteenth century; when automation and dynamos transformed factories, mills, and all manner of transportation in the late nineteenth century; and when data processors transformed the back offices of large corporations beginning in the 1950s. In each era, businesses found ways to substitute capital machinery for human labor—and capital expenditures proved, in economic terms, more attractive than labor. In some cases, work environments once populated by hundreds or thousands of people became work places populated by a few people managing large numbers of machines. In other cases, new roles for people emerged in the work place, especially in those businesses where the uniquely human attributes of people—such as creativity, problem-solving abilities, and interpersonal skills—translated into economic value.

For decades of modern business history, companies have interacted with their customers and markets predominantly through people in frontline service or managerial positions. The rise of the Internet and ubiquitous networks as platforms for commerce in the 1990s created the first glimpse of a different reality—one that put machines to work in frontline positions to manage transactions, interactions, and, ultimately, customer relationships. That revolution was limited in its impact, because it unfolded online. Even with e-commerce revenues set to achieve a significant proportion of retail volume around the world—in the United States, it's projected to pass 5 percent in the next few years—online sales are still a relatively minor event in the greater scheme of commerce. We consumers are analog creatures; we live most of our lives offline, in a world composed of atoms, not bits. Online commerce is a phenomenon that's happened on the periphery of our reality, not at its center.

Now, however, we are entering an age when many of the implications of the networked world are permeating the physical one. Ubiquitous and intelligent networks are reaching into offline realms. Smart and proliferating devices, which allow us to access those networks, are becoming part of our everyday world. The result is that Web screens anchored to bulky personal computers are not the only way in which the intelligence of machines is impinging on our experience of life. Now, a vast array of anchored and mobile devices, in mind-expanding and ever-mutating variety, is positioned to connect us continuously to global networks. That same array of devices and networks makes it possible for companies to relate to customers and interact with markets in radically new ways.

If you stop to think in these terms, there is much around us that is strangely different and even discontinuous from recent commercial reality. Every time an airport kiosk issues a boarding pass, it's playing a role in a dramatic substitution of frontline machine labor for what was, until recently, the exclusive province of human effort. Every time a pharmacy refills a prescription using an automated voice-response system, it's using a machine to assume a critical front-office task once performed by the pharmacist in the store. When you leave a parking garage and use a talking vending machine to accept payment and validate your ticket, the transaction is one that, until recently, required clerks sitting at service counters to execute.

Each one of these examples reflects a facet of the front-office revolution unfolding today—the substitution of machine for human labor in the physical world of business. Automation has come to services. Devices such as kiosks, interactive voice-response units, Web sites, ATMs, and sophisticated vending machines are driving down the costs of customer interactions even as they enable more satisfying customer experiences. While automation of transactional services is not new, machines deployed on the front lines today have reached a threshold of intelligence, interactivity, and emotional appeal that is unprecedented; these attributes, when combined with networking, qualify such machines to manage human relationships with more sophistication than ever before. And the cost compression opportunities of such automation are as dramatic as any in the annals of reengineering associated with the back-office automation of two decades ago—or in the industrial automation movement that began over a hundred and fifty years ago.

At the same time, it's become commonplace to pick up the phone seeking to contact a company and discover that you've reached a call center in a far-flung location. Decades ago, telecommunications and financial services firms first sought to lower their costs of handling service calls by shifting their call centers from locations proximal to headquarters to ones in remote Midwestern states like South Dakota, then to offshore locations like Ireland, where labor rates were lower. With plummeting costs of long-distance communications, this kind of labor-rate arbitrage made sense. Often, those workers in remote locations were friendlier and spoke clearer English than their counterparts closer to home. Today, the fact of ubiquitous communications networks has created more dramatic outcomes, with professional as well as clerical service positions increasingly displaced to remote locations such as China, India, and the Philippines. In each case, corporate employers have concluded that they can hire better talent at lower cost, improving the economics of their businesses while providing a better quality of interaction for customers.

Phenomena such as outsourcing and off-shoring, like automation in services, are part of the same story that's driven by the confluence of these two trends—ubiquitous networks and smart devices. These trends are the underlying themes of this book. We argue that the forces associated with such evolving technologies are fomenting a radical reconfiguration in how the world's leading businesses go to market. Networks create the flexibility in frontline service positions to deploy human talent that is physically proximal as well as geographically remote. Smart devices create the flexibility to interact with customers using machines as well as people. In short, networks enable *displacement* of service roles and functions; and devices enable their *automation*. The possibilities represented by displacement and automation, in turn, throw into question how just about every company competes today. Why? The opportunities for radical gains in efficiency and effectiveness related to how companies manage interactions and relationships with customers can enhance both enterprise economics and the differentiation of a company's offerings and brands.

Of course, just as people who are remote can serve customers by connecting across great distances, machines can do the same. That was one lesson from the Internet revolution. The "death of distance" meant that remote Web servers could process transactions, for example, for e-commerce customers around the world, potentially transforming local businesses into franchises with national or global reach. As these dynamics fold over onto the physical world, they enable both people and machines to serve customers either in direct proximity or from remote locations. And the fact of connectivity itself means that people and machines can collaborate in new ways, too. Hence the power of a skilled call center representative, who can deliver over a phone line human warmth and empathy with database-driven precision in tailored services.

To apply these ideas about networks and devices, and the displacement and automation they enable, to a wholesale rethinking of how companies relate to customers and markets is what we call *front-office reengineering*. It is, we believe, how smart enterprises and their managers will compete for advantage now and in the future.

The reengineered front office is composed of three varieties of *service interfaces*. We define a service interface as a front-office element of operations that mediates interactions and relationships between a company and its customers. Such service interfaces take shape according to three basic archetypes—*people-dominant, machine-dominant*, and *hybrids* of people and machines. We think of an interaction with a waiter as a people-dominant service interface (even if it is supported by computerized ordering systems). We think of a vending machine or a Web site as a machine-dominant service interface (even if it is supported by staffs for maintenance and development). And we think of a call center representative, who cannot perform his or her job without access to phone lines and database systems, as a hybrid service interface.

As companies aim to improve the efficiency and effectiveness of their interaction and relationship management operations, senior executives and managers must ask several critical questions of themselves and their companies:

- Does each service interface perform its functions optimally, or could our company do better by deploying people in place of machines or machines in place of people?

- Does each service interface perform its functions optimally, or could we do better by deploying people in collaboration with machines or machines in collaboration with people?

- Does each service interface perform its role optimally, or could we do better by deploying interfaces remotely if they are proximal, or proximally if they are remote?

In answering these questions, managers will inevitably recognize that some previously deployed interfaces have become superfluous in the reengineered operation. They may conclude that other interfaces are missing. These realizations stem from a fundamental insight that lies at the heart of what we argue in this book. While the vast majority of companies have sunk enormous investments into deploying broad arrays of service interfaces with customers, most do not, in fact, manage their interfaces—regardless of how well each interface is configured—in the context of what we call *interface systems*. When such systems function properly, they represent not a portfolio of uncoordinated touch-points or connections between companies and customers, but rather a unified and unique *interface capability* that manages relationships in integrated and seamless ways. When realized successfully, such interface capability drives down the cost of managing each customer interaction while driving up the quality of interaction. How? By gaining new operating leverage (in both costs and revenues) that ubiquitous networks and smart devices make possible.

We are talking about a radical reconfiguration of work and a radical rethinking of corporate strategy. It's a fact that the vast majority of human beings in industrialized countries are consumed every day by jobs pertaining in some fashion to interaction or relationship management with customers. Leave aside the fact that the service sector itself accounts for nearly nine out of every ten jobs. To change how service work is organized and performed is tantamount to an industrial revolution. Indeed, that revolution—a revolution in services that technology makes possible—is the ultimate subject of this book.

SUMMARY OF THE BOOK

In chapter 1, we examine the new realities of business that have made the quality of customer interactions and relationships the next frontier of competitive advantage. In so doing, we explore the limitations of competing on product and service offerings alone, which have given rise in recent years to a focus on customer experience. We argue that the most actionable approach to managing customer experience centers on a company's service interfaces, because those interfaces are how companies determine the quality of their interactions and relationships with customers. Managers must open their minds to innovative configurations of people and machines to compose interface systems in optimal ways. If they do, the payoffs in productivity gains can be enormous and we examine the literature supporting this position. Finally, we conclude the chapter with a set of principles that guide our understanding of the front-office revolution.

In chapter 2, we explain why the front-office revolution is unfolding now. We explore in depth the trends in technology evolution related to networks and devices that make this a time of unique opportunity for establishing interface-based strategic advantage. While no one trend is new, in combination they create a kind of a threshold effect. The trends are: the proliferation of smart devices, the rising intelligence and interactivity of those devices, the capacity of such devices to appeal on increasingly emotional dimensions, and the synaptic connectivity that links such devices to other devices and networks. These trends result in new possibilities for the roles technology can play in managing customer interactions for companies. In persuasively adopting customer relationship management roles, machines have come of age—and joined the work force, this time in the front office.

In chapter 3, we contrast the reengineering revolution of the 1980s with the front-office reengineering revolution unfolding today. Front-office reengineering involves the radical redefinition of front-office labor in light of the contributions of machines and machine-driven processes. These new roles for machines result from the four trends outlined in chapter 2. In this chapter, we deal with the economic incentives to substitute capital equipment for human labor on the front lines, both for efficiency (lower costs of delivering a customer interaction or relationship) and effectiveness (better quality of customer interaction or relationship management). In our view, front-office reengineering is ultimately concerned with productivity gains at the enterprise level, as measured in reduced costs and increased revenues.

In the next three chapters, we take a close look at the building blocks of interface systems, outlining our three interface types or archetypes. In chapter 4, we examine the traditional interface through which companies have delivered services throughout history—the pure human interface involving people enabled by people in the interaction and relationship management functions of the front office. In chapter 5, we examine the automated interface that has begun to appear as more and more smart devices are tied to networks and become compelling interfaces along physical, cognitive, emotional, and synaptic dimensions, for connecting customers and companies. In chapter 6, we examine two versions of the hybrid interface, people enabled by machines and machines enabled by people. Our intent here is to examine the ways in which humans and machines may collaborate in the front-line work force, and how the combination of people and technology can prove powerful in compressing costs while increasing the quality of customer interactions and relationships.

In chapter 7, we examine a variety of interface systems and what makes some successful and others not. These systems represent combinations of the foregoing interface archetypes, with the added complexity of multiple layers of people and technology creating even more complex, multilevel interfaces. We focus in particular on one company that's orchestrated a world-class interface system using people and machines in a variety of innovative roles—and created a truly breakthrough business as a result. That company is QVC, the leader in TV home shopping and a retailer with some of the highest levels of satisfaction and loyalty, not to mention profitability, in the business.

In chapter 8, we provide an assessment tool for deploying interface systems and optimizing portfolios of interfaces already deployed. This methodology stresses that interface systems have two constituencies—the internal (employees) and the external (customers)—because internal and external interfaces are intimately interrelated and companies must deploy them with this interrelationship in mind. The process of front-office reengineering entails trade-offs in relationship management between efficiency and effectiveness. For every business, the trade-offs drive different forms of optimization. Getting it right enables corporations to manage customer interactions and relationships in ways that create long-term, sustainable competitive advantage.

Best Face
Forward

1

Interfaces As the Next Frontier of Competitive Advantage

In this book, we argue that where competition is overwhelmingly intense and where products and services become commodities overnight, the only lasting competitive advantage will derive from superior interface capability—enabled by a reconfigured front office that takes advantage of the capabilities of both machines and people. Few business sectors can avert intense rivalry and commoditization, so companies must create interfaces with customers and markets that are at once more *effective* (yield a better quality customer interaction) and more *efficient* (incent a better interaction at lower cost per interaction) to create and sustain true competitive advantage. This model is the essence of a new productivity revolution. Its possibilities are fueled in large measure by technology. Machines are poised to play central and sophisticated relationship management roles on the front lines of business. As front-office machines have proven increasingly capable of managing interactions and relationships with customers and markets, they are driving corporations to change how they relate to, and interact with, the world. The quality of interactions with customers—and the customer experiences that result from those interactions—is rapidly becoming the sole remaining basis of competitive advantage.

In espousing this position, we are focusing on the role of services in business as a key competitive weapon. There is a long-standing belief that automation would never drive productivity gains in services as it did in manufacturing, because the delivery of services depended on human labor. Increasingly, while employees continue to interact with customers on a company's front lines, machines are working alongside or replacing them altogether. Either way, technology is managing interactions and relationships with customers and markets as never before.

At the turn of the nineteenth century, a new age of economic activity dawned with the invention of large-scale enterprise and managerial capitalism. The business innovations that gave rise to modern capitalism resulted from revolutions in transportation, communications, and production; automation of industrial processes became feasible by harnessing first steam and then electric power to drive new forms of mechanized production. To realize their visions of mass production and mass markets, entrepreneurs recruited machines into the work force. Innovative machines went to work in operations as diverse as textile mills and auto assembly plants, ultimately moving assembly lines and automated production processes, at times making human workers more productive or replacing them altogether. Plant-floor productivity gains were realized through new management techniques, such as Frederick Winslow Taylor's time-and-motion studies, which drove laborers to work most efficiently by mimicking the tools they used.

Around 1800, roughly 80 percent of the work force in the United States toiled in agriculture. By 1900, that proportion had fallen to 40 percent, and it amounts to less than 3 percent today.[1] The shift occurred partly as a result of machines replacing human workers in farming operations, and partly as a function of the enormous opportunities afforded workers by newly formed large-scale enterprises. Workers streamed from the countryside into large cities throughout the nineteenth and twentieth centuries seeking industrial employment. Of course, the vast proportion of the work force today no longer works in agriculture or industry; we live in an economy that creates its output largely in services, which account for over 90 percent of U.S. employment, compared to less than 10 percent in manufacturing.[2] The redistribution of the work force has accelerated in just the last few decades. In the early 1900s, three out of ten U.S. workers were employed in the service sector. In the 1950s, this figure grew to five out of ten—and, since the early 1990s, it has risen to eight out of ten.[3] This trend reflects the way in which the U.S. economy, in particular, generates domestic output. According to the Organisation for Economic Co-operation and Development, the service sector accounts for more than 80 percent of annual economic output in the United States on a gross domestic product basis. Few developed economies in the world have output from services that's less than 65 percent.[4]

But it's not merely an issue of service businesses having supplanted industrial concerns as the major drivers of wealth creation in our economy. Service has also become a critical strategic weapon in helping companies establish and sustain competitive advantage.

WHY SERVICE MATTERS

Following the shift from agricultural and industrial work to services, something more profound has transpired. Service has become the primary way that firms in rapidly commoditizing markets establish and sustain advantage. This is why the vast majority of jobs involve service functions, including those in manufacturing companies. Employees in service positions account for the majority of jobs in service sector businesses and a large portion of jobs in manufacturing and agricultural companies.[5] Here's why: In an economic context where companies compete on the quality of interactions and relationships with customers, labor flows to those tasks that deliver the highest economic returns on human talent. If the management of relationships between companies and customers creates the greatest degrees of advantage, service positions become the magnet for a growing head count—and this is exactly what has happened.

Recent management literature has suggested that companies must now compete on customer experience, an integrated bundle of products, services, and information.[6] Such arguments tend to view customer experience as an end in itself. In so doing, they admonish managers to use customer experience as a competitive weapon to render what they sell memorable, engaging, and even life-changing or transformational for customers.[7] In our view, many of these ideas are rather extreme. Every business is not a stage, nor is every company interaction an opportunity to share an experiential universe with one another at any time. "Realizing our dreams becomes a driving force" does not hold true for customers in every commercial interaction.[8] First, customers do not want every interaction with a company to constitute a life-altering experience. Depending on the customer and the purchase occasion, what constitutes an appropriate experience may vary wildly. At a fine restaurant, for example, interactions that create memorable or transformative experiences are winners; at a fast-food restaurant, interactions that are quick and routine (literally unmemorable) carry the day. Second, managers cannot conceive of customer experience as an unalloyed good that exists in a world separate from enterprise economics, as much of the literature implies. The management levers required to deliver appropriate and targeted experiences are the points of connection that link companies and customers—what we call *service interfaces*. But managers cannot force a segment of customers to have a specific experience, since each customer's experience is personal and potentially unique. Managers can, however, deploy interfaces that enable certain kinds of relationships between companies and their customers, creating experiences that fit with a company's market position and brand.

Like everything else in business, managing customer interactions and relationships involves trade-offs, economic and otherwise. While experience happens in customers' minds, service interfaces are what managers directly determine and control. To get interfaces right, managers must configure and deploy them based on an analytic understanding of customer needs and desires, taking into account customer segments and purchase occasions as well as competitive offerings. The concept of customer experience should have everything to do with the customer's objectives when interacting with a company and vice versa; and what any one company actually does for its customers is inseparable from its revenues and the margin dynamics of its business.

To define our terms, *interactions* refer to the behaviors of customers when they are engaging with companies; *relationships* refer to how customers come to view those companies cognitively and emotionally over time. Relationships are a way of describing the evolving intellectual and emotional context in which customers' interactions with companies take place. Together, interactions and relationships generate customer experience, which, in turn, influences how future customer behaviors and attitudes unfold (see figure 1-1). Smart managers actually work backward from the appropriate customer experiences they wish to deliver, to the interactions and relationships that shape those experiences, to the configuration of the interfaces and interface systems that will successfully mediate those relationships. Why is this important? Increasingly, companies differentiate their offerings not by *what* they sell but by *how* they sell it.

THE LABOR SCARCITY PARADOX

If you accept these arguments, then it's clear why corporations today are facing an arresting paradox. On one hand, unemploy-

FIGURE 1-1

Construct for Experience

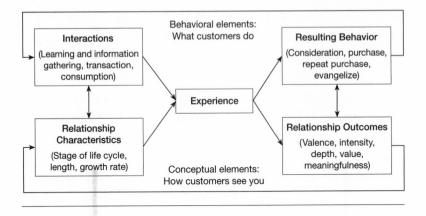

ment levels have been running higher than in several decades in many industrialized countries, and there's widespread concern about both working class and professional jobs going offshore or falling prey to automation.[9] On the other hand, companies decry the lack of skilled or qualified labor to meet their needs in domestic markets and abroad. Call it the tyranny of the 80/20 rule—20 percent of the work force delivers 80 percent of the value—that makes for a chronic perceived shortage of able people in a world of unemployment. The fact is companies are suffering a shortage of affordable, appropriately skilled talent, which will only be exacerbated by the retirement of the baby boomer generation.[10] Why? Because in a work force where the vast majority of people on corporate payrolls are responsible for managing interactions with customers, the desired attributes of the work force have changed. Positions that deal with customer interactions in many industries are either white-collar or require similar skills, demanding employees who have well-honed interpersonal abilities, are literate and numerate, can communicate effectively through multiple channels, and possess analytic capabilities and emotional intelligence.[11] These requirements have driven corporate America to spend $50 billion a year on the remedial education of workers.[12] Qualified personnel with the talent for relationship management roles are difficult to come by—and, when businesses find workers in their local markets, they're often costly to employ because, regardless of the unemployment numbers, they're in perpetually short supply.

The convergence of how companies compete, the dynamics of the labor force, and the evolving capabilities of machines set today's business world at a watershed. In the context of talent scarcity for frontline positions, machines are emerging as viable alternatives— not merely to process transactions, like ATMs, but to manage inter- actions in sophisticated ways, like some interactive voice-response units, in-store kiosks, and Web sites. The celebrated twentieth- century French historian Fernand Braudel observed that in the ancient city of Alexandria in Egypt scientists, physicians, and engi- neers were extraordinarily advanced: They had figured out that the earth revolved around the sun, identified the brain as the seat of thought in the human body, and worked out the principles of steam power.[13] All of those discoveries and advancements took place two centuries before the Common Era—nearly two thousand years before the Enlightenment. But there was no industrial revolution at the time: Alexandria failed to automate physical work because Alexandria had slaves. Scholars of the American Civil War hypoth- esize that the Northern states made the transition from an agrar- ian to an industrial economy long before the South, because the latter had abundant agricultural resources and slave labor in the first half of the nineteenth century. Current analysis of Arab states in the Middle East suggests that the vast wealth in petroleum re- serves has enabled those kingdoms to outsource industrial labor, thereby sustaining Islamic resistance to modern values of capital- ism and democracy.[14]

A scarcity of labor, combined with new technological possibili- ties, creates the dynamics that drive industrial revolutions, wherein entire economies, led by innovative companies, embrace new ways to augment human effort. In such revolutions, we discover new produc- tivity frontiers. For example, the stagnation in customer satisfaction ratings over the past decade can be attributed to the shortage of frontline or relationship-management talent, yielding opportunities to supplement labor with technology. Even as product quality in industries such as automotive and electronics has risen, prices have held steady or fallen; the American Customer Satisfaction Index (ACSI) declined each consecutive year from its inception in 1994 through 1997. It has remained below 1994 levels ever since (see figure 1-2).[15] To call this steadily growing dissatisfaction a predicament is an understatement, but it makes sense in light of the operating and economic challenges involved.

Consumers—endowed with greater choice and transparency in pricing and markets—are demanding more. Just as companies desperately need new and better ways to interact with customers beyond the products and services they sell, their frontline work forces are least suited to getting the job done, let alone to nurturing relationships that might engender loyalty.

FIGURE 1-2

Stagnation of Customer Satisfaction

Source: American Customer Satisfaction Index, "Quarterly National Scores American Customer Satisfaction Index, Q3 1994 to Q1 2004," ACSI Web site (<http://www.theacsi.org/national_scores.htm>).

The Rise of the (New) Machines

In response, innovative companies are recruiting machines into the work force, not just in the back office or on the plant floor, but on the front lines. These new machines help companies go to market, manage customer interactions, and establish commercial relationships. To drive productivity gains, managers have given these machines prominent roles in the once human-dominated world of relationship management. A combination of smart devices linked to intelligent networks—Web sites in retailing, ATMs in banking, touch-screen kiosks in shopping, drive-thru windows in fast food, e-ticketing machines in airports, and slot machines in casinos, to name just a few—enables businesses to deploy machines in managing customer relationships in radically new ways, either to augment human labor or to replace it altogether. The promise of domestic service robots, which was once the stuff of science fiction, is coming of age. Products such as Roomba, the robotic vacuum cleaner that resembles a small flying saucer, sold over 200,000 units during its first year of sales.[16] Sony's AIBO robotic dog has become a cult fascination among Japanese and American owners who have fallen in love with their intelligent, if artificial, pets. And corporations including Hitachi, Honda, and Toyota have introduced bipedal robots as exemplars of a new technology platform with the economic promise that automobiles and personal computers (PCs) had.[17] These machines can serve customers by delivering more than functional value; they can often elicit responses from people that are emotionally powerful.

Put together, these examples represent a striking reversal of business realities from those of the late nineteenth century. Our world's new machines have entered the work force in an analogous process of capital-for-labor substitution; but while the nineteenth-century's new machines interacted with industrial goods and production processes, our new machines interact with people and social processes. This reversal represents a new mandate for business managers that is the inverse of the managerial challenge in the preceding century.

Today, leading businesses don't need people in the work force to behave more like machines—they need machines to behave more like people. Just as some machines in industry proved more skilled than humans in performing certain tasks while incurring lower costs, today's frontline machines are driving a parallel revolution of effectiveness and efficiency. What we call the *front-office reengineering revolution* is utterly changing how companies organize work, relate to customers and markets, and establish competitive advantage. Let's make an important distinction: We define the *front office* as the realm where the delivery of services and the management of customer and market relationships take place, including marketing, sales, channel and technology management, and customer service. We define the *back office* as the realm where manufacturing or production occurs, along with supporting functions such as administration, human resources, finance, and supplier relations. Consider Fidelity Investments. Its marketing and sales happen in the front office, and its management of funds in the back office—so much so that Fidelity managers refer to fund management as the "factory" and the rest of the organization as sales.

Necessity dictates that businesses hire and deploy machines in the front office, not just in the plant. Never before have managers had access to technology that could credibly play front-office roles. Never before has large-scale enterprise possessed the technological infrastructures or networks—resulting from decades of sustained investment in information technology in major corporations—to deploy front-office machines.[18] Never before have sophisticated companies understood the importance of placing technology investments in their appropriate organizational contexts, involving the interplay of technology with managerial capabilities, organizational design, and work processes.[19] Leading companies have begun to explore how to deploy machines not only in blue-collar roles, performing menial physical or computational tasks in the back office, but also in white-collar roles, managing interactions between companies and customers in the front office.

This book is an examination of that revolution—an industrial revolution and a reengineering revolution. Our hypothesis is: *How companies structure and manage their entire system of interactions and relationships with customers and markets will substantially determine their ability to establish and sustain competitive advantage.* The primary functions of the front office—with its emphasis on the elements of customer interaction and relationship management, such as marketing, sales, service, and brand—exist to support what we call the *interface* or, in most cases, the *interface system* of a business (see the glossary at the back of the book). Customers experience this set of interfaces—some human, some machine, and some combinations of human and machine; some proximal and some remote—when interacting with a company.

There is a confluence of mandates here. On one hand, businesses must sustain customer-satisfying and loyalty-inducing interfaces to win in a world that competes more than ever on the quality of customer interactions and relationships. On the other hand, technologies related to smart devices and intelligent networks create new possibilities for how companies can configure and deploy their service interfaces. In this confluence lies the key to the future of competitive advantage. New competitive realities and technologies have converged to drive businesses to adopt more favorable operating models that will enable them to orchestrate better customer relationships at lower unit cost per relationship—it's a new frontier of efficiency *and* effectiveness.

Cost As a Driving Force

The economic drivers of this revolution are undeniably compelling. Consider the world of customer service. Service industry data suggest that it costs the average company $9.50 to respond to a telephone inquiry from a customer. Addressing the same customer needs live by e-mail costs $9.00. Meeting those needs in online text chat involving a live person, where that person can handle several customers at once, costs $5.00. Handling the interaction using e-mail by a live person with automated assists or macros costs $2.50. But if the interaction is fully automated, the savings are dramatic. An interactive voice-response unit can handle the inquiry for just $1.10, if it does not default to a human being; a Web site can handle it for $0.50; and an automated e-mail response unit, if suitable, can handle it for just $0.25. When you consider that the world's call centers were handing 26 billion call minutes a month by mid-2003—with projections of 35 billion call minutes a month by 2007—it's hard to overestimate the economic impact of this kind of front-office automation.[20] Similarly, consider the case of robots: According to *World Robotics 2003*, a report by the International Federation of Robotics and the United Nations, if you index the price of robots and the cost of human labor with 1990 as the baseline, the robot price index has decreased from 100 to 36.9 (or to 18.5 if you adjust for the higher quality of today's robots) while the index for human labor compensation has risen from 100 to over 151 (see figure 1-3).

This dramatic cost-savings is why many sophisticated e-commerce companies manage a critical aspect of their cost structure through a metric that tracks customer contacts per order: The more transactions that a Web site can complete on a fully self-service basis, with no human contact, the lower the overall costs of operating the business. And, contrary to popular belief, this is not a negative outcome from the customer's point of view. For example, Amazon.com has some of the highest customer satisfaction ratings of any retailer offline or online, yet its site is carefully engineered to discourage human contact during the shopping process; the company no longer runs massive call centers, as it once did, and it responds to customer inquiries almost exclusively by inbound and outbound e-mail. As a result, Amazon's customer service costs run less than 1 percent of sales, while other online retailers operate with ratios that are nearly ten times that level. Of course, there are other advantages to interactions free of human assistance: Customers using automated systems feel more in control, can proceed at their own pace, experience a greater degree of privacy, and may appreciate the absence of inattentiveness or a poor attitude from human customer-service representatives.

To see how machines are changing front-office operations today, consider a few examples across the diverse sectors of fast food, financial services, transportation, health care, and logistics:

FIGURE 1-3

Price Index of Industrial Robots Versus Human Labor

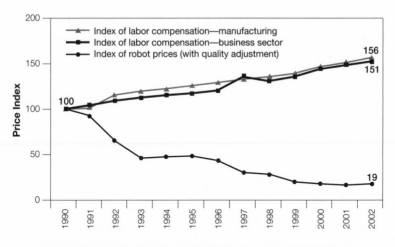

Source: International Federation of Robotics and the United Nations, *World Robotics 2003* (Geneva: United Nations, 2003), 6.

- *Krispy Kreme* uses an Internet application to link its stores to coordinate supply ordering and to monitor overstocks. While this innovation reduces drudgery for employees, it also frees them up to spend more time with customers. Problem orders have dropped 90 percent, from 26,000 to fewer than 3,000 over three years, and only ten district managers are required to manage a 320-store national chain, as opposed to 144 three years earlier.[21]

- *Charles Schwab & Co.* offers advice to high-net-worth individuals without having recruited a department of financial analysts. With a $20 million investment, it created Schwab Equity Ratings, an automated online service that provides buy and sell recommendations for roughly three thousand equities. With computers making stock picks that Schwab claims equal the quality of those made by investment professionals, it can offer sound advice at a substantially lower cost.[22]

- *Progressive Insurance* furnishes its thirty thousand agents with access to a Web site that provides price quotes for policies and online loan applications. Agents who use the site are 40 percent more efficient, the site has reduced Progressive's costs to 20 percent of revenue (the industry average is 23 to 25 percent), and its revenues have grown four times faster than the competition's.[23]

- *Northwest Airlines* has installed 755 automated check-in kiosks at 188 airports. Two-thirds of its passengers either use the kiosks at the airport or print boarding passes at home from the airline's Web site, up from 20 percent in 2001. Industry experts estimate that one automated kiosk can replace the work of 2.5 employees, while incurring costs in maintenance and upkeep that are roughly one-fourth the annual expenses for a single full-time employee.[24]

- *Sutter Health* has enhanced its emergency rooms with a $20 million system called eICU, which combines patient monitoring with digital record keeping. The system streams vital statistics and visual monitoring of patients by Web cam to doctors. The result is fewer complications (blood-clot and stress-ulcer rates have fallen from 25 percent and 14 percent, respectively, to practically zero), resulting in better care and shorter stays in the hospital.[25]

- *FedEx* completed a $150-million upgrade of handheld barcode-scanner and order-entry devices carried by its delivery personnel, which enables them to provide better service and more accurate information to customers. The devices save at least $20 million a year in operations, by shaving ten seconds off each pickup.[26]

Humanizing the Work Place

If those examples strike you as a vision of a cold and impersonal face of business, they're not. We believe that this revolution will deliver the opposite effect. Yes, machines will displace some jobs, but they will create others. You needn't look far in the consumer world to understand that many people in frontline jobs, especially in large-scale enterprise, are engaged in mindless, unfulfilling, and psychologically inhumane work. Whether it's performing repetitive tasks in call centers or demeaning chores in fast-food restaurants, people working in the front offices of most large-scale service enterprises spend their days in anything but enlightened activity. They toil in today's equivalent of Gradgrindian mills—the sweatshops of a postindustrial society. In most frontline service positions, interacting with customers is treated as an afterthought, indulged only if there's time and energy after performing the core service or delivering the product. Front-office machines will change that. Their very existence, like that of steam power in the nineteenth century or data-processing in the twentieth, will require executives and strategists alike to rethink how businesses manage people, configure work processes, and deploy human and machine labor. Paradoxically, front-office machines may humanize the face of business.

We see a world where companies engage customers more reliably than today by means of consistently high-quality interactions and satisfying relationships. In doing so, businesses will create dramatically more-satisfying jobs and work environments for employees. During the waves of industrial revolution in the nineteenth century, automation of the plant floor created a host of industrial nightmares, with workers suffering inhumane conditions and ultimately rising up in violent protests, from the Homestead Strike in 1892 and extending to the General Motors sit-down strike in the 1930s. Only after several decades of plant-floor automation did the human face of business reassert itself and create industrial work environments where human dignity among production machines became a core tenet of management. The division of labor between humans and machines in the front office need not rehearse such dark passages. If managed appropriately, the front-office revolution will bring machines into positions that they perform well—

and, more important, it will bring people into roles they perform best. This revolution must result in the humane use of employees; business does not employ the vast majority of frontline workers in humane positions today. Far from taking people out of the work place, we believe technology can create more and more ennobling jobs in frontline positions, resulting in better experiences for customers and greater competitive advantage for companies.[27] Just as machines eliminated the hard labor of hauling goods across great distances in the transportation revolution, obviated the need for large numbers of people to toil in fields in the agricultural revolution, eliminated the need for humans to perform repetitive tasks in the manufacturing revolution, and relieved people of mindless number crunching in the data-processing revolution, machines can create a better world for humans in the front-office revolution. Despite the inroads of customer relationship management and sales force automation software, services in general are still relatively untouched by automation. The ultimate outcome depends on the wisdom and creativity of managers who deploy these new technologies. Senior managers leading front-office reengineering can and will determine the effects on employees, customers, and, ultimately, shareholders. A technological nightmare of abusive automated services (i.e., voice-mail hell) may prove easier to deploy and cheaper to operate in the short term, but will destroy companies in the long term by betraying either of the critical human elements of the equation, employees and customers.

PRODUCTIVITY COMES TO SERVICES

The idea that the service sector could wring significant productivity gains from any form of technology or automation is relatively new. For years, economists believed that, due to the centrality of people in service delivery, scale economies were difficult to realize and opportunities for automation were limited. Many have argued that for service businesses, such as professional service firms, diseconomies of scale were the rule. Perspectives of this kind became crystallized in the economics literature as *Baumol's Disease*, a phrase coined by William Baumol, an economist who argued that service sector productivity would lag increasingly behind that of manufacturing because of services' dependency on the human factor.[28] Because services account for the greatest proportion of jobs in the industrialized world, more recent commentary, which suggests that Baumol's Disease might prove curable, has produced a stir—and understandably so. Such a cure would have truly material implications for the enterprise economics of individual firms as well as the output of national economies concentrated in services.[29]

The debate over potential productivity gains in services connects to the question of a payback on IT investment, because IT investment has largely focused on gaining greater efficiency in service-related tasks. On one side are skeptics such as Stephen Roach, chief economist at Morgan Stanley, who has argued since the mid-1980s that the return on IT investment among major corporations has proven limited or nonexistent. Such economists suggest that IT has represented the worst of all worlds; it has raised the cost of doing business while generating no concomitant lift in the creation of value. That's why Robert Solow, another economist in the skeptics' camp, referred in 1987 to the nonexistent returns on IT investment as a "productivity paradox." As he archly characterized the situation in the early boom years of the PC era, "You can see the computer age everywhere but in the productivity statistics."[30]

More recently, positions among economists on this issue have begun to change. Researchers such as Dale Jorgenson of Harvard University and Erik Brynjolfsson of MIT have argued that steady investment in IT over the past two decades has led to productivity gains despite claims to the contrary.[31] Their arguments are based on the notion that enhancements to technology within complex organizations do not translate directly into productivity gains. To realize such benefits, managers must absorb the implications of the new technologies for their businesses—and then make the necessary changes to work processes and organizational design in order to harvest their productivity potential. This camp argues that only in the last few years have corporations across many sectors of the economy begun to act on an understanding of how IT can change the performance characteristics of their businesses. How significant and how generalized might such productivity gains be? Previous industrial transformations—for example, the conversion from steam to electric power—took decades to deliver productivity paybacks, as managers struggled to put the more efficient source of power to work. Similarly, airplane travel took decades from the dawn of flight, including subsequent generations of technology (notably, jet airplane power) and a watershed product innovation (the Boeing 707), to create a market for commercial transport of passengers and cargo.[32] Networks and devices, the twin foci of IT investment, present an even more complex picture. Since IT pervades all sectors, its introduction is, in some sense, analogous to electricity. Like jet-set mobility, it both powers work and creates the potential to radically reshape work processes. As a result, realizing the returns on IT investment is fraught with subtlety and uncertainty. It depends on the redesign of work processes. Whether through outsourcing or off-shoring, integration of automation technology in frontline work, or wholesale automation of frontline labor, seeing the gains is dependent on allocating organizational resources far beyond the realm of IT.

The proponents of an IT productivity payoff identify two critical areas warranting such organizational or business process investments. These IT-related intangible assets are:

- *Organizational capital*, including the configuration of new business processes, work practices, and overall organizational design

- *Human capital*, including worker training or retraining, with a focus on developing higher levels of interpersonal skills, decision-making acumen, and management effectiveness[33]

And, to realize impact, the investment beyond IT is significant. At companies where managers have realized success, researchers

argue that it has proven necessary to match every $1 of IT invest-
ment with up to $9 of investment in IT-related intangible assets.
With such investments in place, productivity gains become both
measurable and material.

A Company-Level Perspective

The idea that it is insufficient to automate company operations
without also reorienting the people and redesigning the processes is the economic underpinning of the reengineering movement and represents a foundation for the arguments in this book.
Realizing the true potential of the front-office opportunity depends on wrestling with far more complexity than the mere acquisition of networks and devices; the payback will depend on
reconfiguring people and organizations to get service interfaces
right. Such perspectives are evident when we examine some of the
many companies that have begun to realize early success in using
technology to reconfigure how they interact with customers.
While we are still early in this unfolding revolution, the data is
persuasive in its weight and implications. Gains attributable to
front-office redesign will represent one of the signal business developments of our time; and we will examine that data in detail in
the chapters to come.

CONCLUSION

In this book, we explore the ways in which front-office machines
will be deployed as well as the elements of new interface forms—
human, machine, and combinations of the two—and their organization into interface systems. We have discussed a number of ideas
that apply.

- *Companies compete more than ever on the quality of their service
  interfaces with customers.* The dominance of the service sector
  in the total economic output of today's economies is part of

our story, but dominance of service-related jobs in companies across all sectors is the headline that matters. The majority of workers today—white-collar or blue-collar, managerial or clerical—focus their activities on managing their companies' interactions and relationships with customers. That's where the greatest leverage for competitive differentiation resides.

- *Nearly all companies already rely on a variety of human and machine interfaces to connect to customers and markets.* While most companies have in recent years allocated such resources, few companies have found ways to manage this proliferation of interfaces strategically. Similarly, few companies manage these interfaces as a system, with an explicit return on investment–based perspective on performance.

- *Customer experience is the result of interactions and relationships that customers have with companies.* Applying a strategic or operational focus directly on customer experience is a recipe for ineffective management. Managers do not control or determine customers' experiences; customers do. In contrast, managers do directly control the service interfaces that mediate customers' interactions and relationships with companies. So managers must deploy these interfaces strategically based on what they sell and on how they sell it.

- *After years of investment in IT within corporations around the world, there is growing evidence that such capital expenditures are beginning to pay off.*[34] The return on IT investment, however, requires that companies adopt radical new approaches to organizational process and design as well as to focused development of human resources. Without such parallel efforts, IT can often raise the cost of doing business without generating any compensatory rewards. Nowhere is this more true than in the deployment of a company's service interfaces, where it can result in too many people and too many machines operating with insufficient coordination and often at cross-purposes, resulting in rising complexity, costs, and customer dissatisfaction.

GUIDEPOSTS FOR THE
FRONT-OFFICE REVOLUTION

This section recapitulates the guiding principles of the front-office revolution introduced in this chapter and detailed in subsequent chapters. As you read through the book, you may find it useful to refer to this summary from time to time.

A New Division of Labor

As this industrial revolution unfolds, machines will increasingly assume both white-collar and blue-collar positions in managing interactions and relationships with customers. People and machines will prove viable in both proximal and remote locations.

How senior executives allocate front-line labor among people and machines, and according to what geographies, will have decisive impacts on their companies' cost structures and growth; their cultures and organizations; and their expressions of brand. Because the sheer numbers of humans affected by this transition are vast, the weight of these decisions transcends the impacts that we might have witnessed during the industrial automation era. For this reason, the wise division of labor between people and machines will become the primary task of strategy and management. Jobs that gain maximum leverage from human capabilities, involving creativity, warmth, and empathetic response, must go to humans. Jobs that gain maximum leverage from machine capabilities, involving repetitive, data-intensive, and transactional tasks, will likely go to machines. Jobs that are better configured remotely will move off-site and, potentially, offshore. The moral imperative, however, must remain clear: Every one of these decisions must be guided not only by the economics of enterprises but also by the ethical goal which is best stated by borrowing a phrase from Norbert Weiner—namely, to assure that we make "human use of human beings."[35]

Training Machines to Act More Like People

At the dawn of industrial automation more than a century ago, it became the goal of many managers to train factory workers to behave like machines. Getting employees on plant floors to adapt their actions to the technologies of industrial production was the key to making lines run faster. Time-and-motion studies became the key to new frontiers of industrial productivity. Today, business competition is arguably won and lost based far more on the quality of customer and market perceptions of companies than on their industrial efficiency. In most industries, we take world-class production capabilities (e.g., pertaining to quality and cost) for granted. In consequence, as machines move into frontline roles, we must train them to act more like people—and to manage interactions with people in increasingly appealing and compelling ways. Because our expectation is that customers, at least, will remain human for the foreseeable future, the reengineered front office must deploy interfaces that appeal to them in aesthetic, intellectual, and emotional ways, whether realized by people, machines, or a combination of both.

Better Jobs for People

One interpretation of the trends in this book is that they represent the death-knell for meaningful or abundant work performed by people. That is not our view. We believe that the choice facing senior executives is clear: We can create better roles for humans as a result of ushering machines into the frontline work force, or we can aspire to a wholesale substitution of machine for human effort. We believe that the argument for substitution is naïve. Consider the predictions made several decades ago regarding the impact of computers on the work place. Futurists intoned predictions of a radically altered work place resulting from the digitization of information: the paperless office. Of course, digitization created great change, but one impact was *not* the elimination of paper; in fact, we use and live with more paper than ever before. An important aspect of how we use paper has changed, however. Unlike three decades ago, we have an alternative display medium for information—the PC. In offices every day, we make choices about the appropriate division of labor between paper and screen. Because of the screen, we use paper for more specialized roles, even if we happen to use more of it. The revolution in services will deliver similar results. Businesses will still employ vast work forces of human beings, but their roles will become more focused and more specialized around the essential humanity of people than ever before.

Technology Is Neutral

A corollary to the preceding principle is a dictum often intoned by technologists—"technology is neutral." This is not an apology or a dodge; it's a fact that machines have (as yet) no moral intent. The entry of machines into the front lines of business is not an option for most companies. In many companies, it's already happened. But how machines affect and alter the work place is an option— their impact represents for most companies a set of significant strategic and cultural choices. Company cultures may go the way of an Orwellian nightmare, or machines can humanize corporate cultures by supporting and amplifying those organizations uniquely human qualities. Nightmare or doomsday scenarios notwithstanding, how new technologies change our corporations is up to us.[36]

Businesses Will Compete Interface System
Versus Interface System

In industrial markets, managers often characterize competition in recent years as having shifted from company versus company to supply chain versus supply chain. Consider Wal-Mart or Dell. No company could successfully challenge either one of them as an isolated entity: A competitive assault would require orchestrating parallel supply chains to match the efficiencies and responsiveness of what those companies have already deployed. Similarly, as the frontier of competitive advantage shifts from specific product and service offerings to the quality of a company's customer interactions and relationships, the importance of interface capability rises. That means that competitive success for most large-scale enterprises will depend on the efficiency and effectiveness of their interface systems vis-à-vis their competitors', as the pursuit of growth demands that companies achieve ever more satisfying experiences among customers and markets.

Interface Systems Are the Ultimate Expression of Brands

There is a time-honored dictum in education: "Every conversation is an opportunity to teach." In the corporate world, one might argue that this expression has its analogue—"Every interface is an opportunity to express a brand." Years ago, managers assumed that brand was uniquely associated with a logo or a tagline, or with advertising and promotion. Today, managers understand that customers' brand perceptions are shaped by every interaction they have with companies, including call centers, salespeople, and direct mail, not to mention competitors' activities in the market. Every interface is a moment of truth for the brand; and every interface system is a potential embodiment of, or detriment to, brand perceptions. A portfolio of interfaces operates successfully only if it's managed in internally consistent and aligned ways. To create distinctive perceptions of brand, managers must coordinate every point at which customers interact with or relate to what their companies do. Interface systems create the context in which brand-consistent interactions must take place.

Companies Must Take Complexity Away from Customers

The goal of interface systems is to manage complex relationships in simple and intuitive ways for customers. Too many interfaces in a system are as deleterious to the quality of a company's customer relationships as too few. As Albert Einstein opined, "Everything must be as simple as possible, but not simpler." Interface systems have a natural tendency to grow more complex over time. Companies either add interfaces or elaborate existing ones in response to customer demand and innovation. Without systems thinking, this process often goes unmanaged, degrading the quality of service interactions for customers while increasing the costs in complexity and lack of integration for companies. The goal must be clear: Even as managers aim to become market-driven, they must configure interface systems in ways that deliver external simplicity even if it creates internal complexity. Such complexity within the company may result in unattractive economics, which in turn may require trade-offs between company and customer goals. That's why a sound interface system, like good strategy, is best produced by choices and trade-offs, not by proliferation and compromise.

Every Interface System Belongs to Two Worlds

Every interface must operate internally for a company's employees even as it interacts externally with a company's customers. After all, for interfaces to work as a system, the people and machines enabling them must become interoperable with one another. For example, a call center representative who has no familiarity with the company's catalog or Web site is unlikely to prove effective in the service of customers or the enterprise. While many companies deploy different interfaces for employees and customers, the logic still applies: Employees and customers ultimately interact in the context of interface systems. As such, neither the external nor the internal realization of an interface system can be considered without the other—they are two sides of the same coin.

Without a doubt, the focus of this book is an unfolding story. The underlying technology driving this revolution is changing from one day to the next. Nonetheless, we can see with relative clarity the outlines of what the next few years in business will bring. As we discuss in this chapter, the evidence suggests that customer interactions and relationships are the next frontier for competitive advantage. The evidence also points to the role that ubiquitous networks and smart devices will play in providing the basis for reengineering the very nature of service work. This reengineering will involve not only a new division of labor between people and machines affecting the vast majority of workers around the world, but also a new design for labor involving proximal and remote roles. The time for action is now. Building a company's interface capability will prove a strategic imperative—perhaps *the* strategic imperative for employees, customers, and shareholders alike. And senior managers who aspire today to own this future can and must develop the interface systems that will support their competitive objectives for decades to come.

2

THE INTERFACE
IMPERATIVE

FROM THE EARLY DAYS of the first industrial revolution, the idea of machines assuming the work of humans—first in agriculture, then in manufacturing—has become a familiar notion. There is strong evidence to suggest that the very concept of unemployment, which is infinitely familiar to us today, only became part of social consciousness at the beginning of the industrial era.[1] Until the mid-nineteenth century, when machines first began to relieve human toil on a widespread basis, the idea of enforced idleness among workers was largely unknown. No agricultural or manufacturing enterprise could ever find too many workers; and no one who wished to work ever had zero options when it came to finding a job. Even so, in that era, one type of work that machines did not touch was the human realm, the locus of interactions and relationships between companies and their customers.

That is one of the dramatic shifts that's unfolding in business today—which is why concerns about chronic levels of unemployment are again heightened. Machines are not merely performing physical tasks in the fields or mechanical tasks in factories or even number-crunching tasks in the corporate back office. Machines are gaining significant influence over how companies project brand images, deliver products and services, and manage sales, service, and customer care. Just as business was incredulous that steam-powered turbines could drive textile mills in the late nineteenth century at lower costs with higher-quality output, many business observers today are unconvinced that machines can manage relationships in credible, customer-satisfying, and loyalty-inducing ways. But it's a fact that machines are beginning, for better or worse, to play or restructure such roles, and are encroaching on a sacred precinct of human activity. It's a giant leap from the steam engine, the cotton gin, the internal combustion engine, the moving assembly line, the vacuum tube, and even the transistor, all of which signified production or back-office technologies. Machines are now relating in meaningful and purposeful ways to human beings.

It's hard to miss the radically altered and evolving role of machines in our lives—how we interact with them, how they interact with us, and how they have become part of our daily lives and business organizations. For example, they now routinely interact with customers in the delivery of services. Think of straightforward electromechanical devices, such as vending machines and coin-operated newspaper boxes, or more complex microprocessor-controlled devices, such as ATMs and interactive voice-response units. In each case, humans—soda jerks, news vendors, bank tellers, and call center representatives—were replaced in a service role by a machine. Simple substitution of machines for people in front-office positions has only proved to be the first wave. After all, elevator buttons replacing elevator operators is an old story; as are dishwashing machines replacing kitchen cleaning staff; freezer and refrigerator appliances replacing visits by the iceman; and home security systems replacing private guards. Even without the use of electronic or electromechanical devices, we have long lived with analogous substitutions of various media for people, such as mannequins in department store windows replacing live models; cutout figures of flight attendants outside airport lounges replacing live greeters; and stylized plywood replicas of Tokyo policemen replacing local police officers at roadside construction sites. These days, such substitutions are commonplace. The difference now, however, is the fusion of such media interfaces with machine capabilities, resulting in machines being both substitutes for human forms and actors performing many once uniquely human roles.

That is what's unfolding in many a reengineered service environment today, wherein a front office of people and machines, in collaboration, manages customer interactions in new ways. Let's revisit those e-ticketing machines at major airports. Most airline service counters use these machines in a hybrid format involving both people and machines, where the machines enable customers to purchase tickets, check themselves in on flights, and, in some cases, select their seat assignments. But the machines don't do everything. They are arrayed along the service counters or clustered out in front. A few frontline service representatives remain behind the counter, even if most of their workstations are gone. In these settings, people are present to deal with the exceptional situations in which machines cannot get the job done—such as physically receiving checked bags, managing complex or problematic ticketing situations, and dealing with delayed or cancelled flights. While the division of labor is relatively straightforward—and the integration of human and machine effort is limited—the outlines of the scheme are clear. Machines deal with repetitive, rote, or standardized tasks that require fast, accurate database-driven responses. Humans deal with unexpected, problematic, or creative challenges that require empathy, interpersonal skills, and an ability to deal with the unexpected. It's a design for work explicitly predicated on maximizing leverage from what people do best and what machines do best.

In most businesses, this amalgam of humans and technology on the front lines is new. Even for the airlines, the experiment in its modern form is only a few years old—though Southwest Airlines introduced vending machines to sell tickets as far back as the 1970s (with the LUV Machine) and Continental Airlines first deployed electronic kiosks in 1995.[2] What's remarkable is the pace at which both companies and customers have adapted to these new possibilities. It's only been twenty-five years since the introduction of the ATM, fifteen years since the advent of voice mail, and less than ten years since the introduction of automated gas pumps. Nonetheless, businesses have been steadily ushering new interface technologies into the work place—reconfiguring working environments to successfully deploy them—and, in most cases, customers have been responding with satisfaction and loyalty. In many settings, machines are providing a higher quality of interaction than the available alternatives in terms of suitable and affordable human talent.

But let us take a step back to understand why and how such interface technology has recently become widespread in the business world in the first place. Then we may begin to understand the ways such technologies will become essential instruments of strategy in establishing and sustaining competitive advantage. As we proceed, please bear the following questions in mind:

1. What is it about this period in the development of business and technology that is yielding such surprising and unprecedented marriages of human and machine labor?

2. Why is a shortage of qualified talent driving business to embrace technology now in radically new ways?

3. In what ways has this amalgamated labor model begun to alter the dynamics of how companies compete?

4. Why is technology now at the forefront of customer relationship management?

Those are pressing questions for anyone who cares about the future of business, let alone that of the human condition.

The Era of Total Commoditization

We know that practically no businesses on earth can survive by leaving well enough alone when it comes to products and services—or what we call their *offerings*. Markets move too quickly, offerings evolve too fast, and the table stakes to play the game keep rising. Several years ago consumer electronics executives in Taipei developed the habit of using the English phrase "three-six-one" to refer to the competitive dynamics of their business. What they meant was three months to create a feature, function, and price configuration that represented a differentiated offering in consumer markets; six months to harvest the margin afforded by that differentiation; and one month to liquidate excess inventory after the offering had become a commodity. A ten-month product life-cycle![3] We view such dynamics that define our economic reality as *the era of total commoditization*.

In a three-six-one world, the new competitive frontier exists inevitably at the customer's relationship with a company and its brand. While offering-based advantage may prove viable from time to time, it no longer represents a sustainable or lasting strategy. Competition in most markets is simply too intense. A customer's perception of a company or its brand is largely determined by the interfaces that the company deploys to manage its customer interactions and relationships. We define such *interfaces* as any place at which a company seeks to manage a relationship with its customers, whether through people, technology, or some combination of the two. Historically, a company's interfaces have consisted of sales, marketing and brand communications, and employees in service positions. As the frontier of competition has shifted from the *what* (the content of the offering itself) to the *how* (the context in which the offering is acquired, experienced, and consumed), companies have focused increasingly on optimizing the customer and market interfaces they manage. This explains, as we noted earlier, why the vast majority of jobs in the industrialized world are, in effect, roles in managing interactions and relationships between companies and their customers. In an economic context where companies compete ultimately not on the quality of their offerings but on the quality of their relationships, this distribution makes sense. The evidence from business history illustrates this point: Over the last hundred years, we have witnessed an inexorable shift in power, the culmination of which is today's focus on customer relationships. We have seen the claim to economic and industrial power move from suppliers and makers of basic raw and finished goods in the 1890s through the 1920s; to assemblers and providers of mass-market products and services in the 1930s through the 1950s; to distributors in the 1960s; to wholesalers in the 1970s; to retailers in the 1980s; and to customers and consumer markets in the 1990s.[4]

The rise of customer power in our era in no way eclipses the importance of creating raw and finished goods or of performing distribution, wholesale, and retail activities—the basic activities that compose elements of the traditional value-chain model of business. It simply suggests that creating those goods and performing those activities have become the minimum entry requirement to play the game. That's why interfaces matter. Getting them right is no longer an option; the value-chain model, which included sales and marketing as afterthoughts as compared to more critical industrial activities, no longer fully reflects business realities. Creating advantage through interactions and relationships has become *the* strategic imperative for companies and their managements—or what we call *the interface imperative*.

The Rise of Customer Apathy

In large-scale enterprise, the challenges involved in orchestrating optimal interactions and relationships with customers consistently over time and distance are not trivial. Indeed, getting tens of thousands of frontline sales and service personnel at hundreds of service locations to act in ways that align with a company's brand intent and market positioning is costly and can prove overwhelming to management. The burdens of scale and complexity tend to result in underwhelming interactions for customers. Either they have inadequate access to qualified company personnel (skilled people are needed, but they're unavailable) or adequate access to unqualified company personnel (unskilled people are available, but they can't get the job done); either way, the customer loses. As the reach of large-scale enterprise has grown and the imperative to compete on interactions and relationships has risen, it's ironic, if not surprising, that measures such as the American Customer Satisfaction Index, as we have seen, tell a story of persistent business dysfunction. Fully 45 percent of senior executives in a recent *BusinessWeek* survey of major corporations believed their companies did not deserve, and had not earned, their customers' loyalty.[5]

Business trade books from years gone by are filled with references to the "service economy," the "experience economy," the "entertainment economy," and the "digital economy" as business scholars and consultants have attempted to pinpoint the essential challenges of doing business in today's world.[6] With the benefit of hindsight, it's clear that each of those arguments had value, but that each addressed a symptom, not the cause. Business has changed. The frontier of competition has shifted. A new center of gravity has become a reality. It's not that business literally competes on attributes such as experience, service, entertainment, or even technology. All business is based on an exchange relationship between a company and its customers; if managed appropriately, such exchange creates value and establishes a company's economic viability. The interactions or relationships that customers have with companies are the context of that exchange. The context determines, in large measure, how companies are positioned strategically and how they compete. Optimizing interactions and relationships is the essence of the competitive challenge in a commoditizing era.

The Decline of Offering-Based Advantage

What has brought us to a threshold in the previously gradual encroachment of machines upon traditionally human tasks of delivering services, managing relationships, and interacting with customers? The answers lie in three macroeconomic trends: product life-cycle acceleration, overcapacity, and margin compression.

Accelerating product and service life cycles

Product life-cycle acceleration may be more pronounced in microprocessor-related products than in other sectors of the economy, hence the cautionary tale of the Taiwanese consumer electronics sector. But it's fair to say that every business these days is competing to survive in a three-six-one world. The staying power of any given product as a differentiated offering is limited, as is the staying power of any potentially unique approach to a particular service business. The world of fashion-forward apparel dramatically exemplifies product and service life-cycle compression. Only a few years ago, the state of the art in apparel retailing was Gap, which famously delivered new merchandise to its stores six times, or seasons, a year. Today, Gap is a struggling brand fighting to make its way back to viability and profitability under new leadership. Competitors have attacked its market with a Gap formula on steroids. The Swedish retailer Hennes & Mauritz, or H&M, tracks and responds to fashion trends with extraordinary speed. H&M introduces new merchandise in its stores on average every three weeks, which means that it effectively operates seventeen seasons a year. The smaller Spanish fashion retailer Zara moves even faster. The total time elapsed from a design concept to apparel on store shelves at Zara is two weeks, resulting in an effective twenty-six seasons a year.[7] Of course, no organization could achieve such cycle times without making use of sophisticated information technology throughout its operations. The presence of such technology—and the lightning-fast pace of its evolution—has made new approaches to fashion-forward retailing possible. H&M and Zara cater to demanding consumers who have the power, and they require a three-six-one pace of innovation just to stay in the game.

Supply outstripping demand

In fundamental ways, our world is an extreme reversal of the postwar economy of just a half-century ago. In the 1950s, there was rising affluence among consumers as the baby boom generation established itself, had families, and grew increasingly affluent. The challenge for businesses at that time was not finding customers to buy goods, but generating sufficient goods to meet the ever-rising tide of customer demands. Today, overcapacity exists in telecommunications, chemicals, automobiles, pharmaceuticals, and consumer packaged goods, among many other sectors. Consider that in 2002 U.S. automakers operated at capacity for combined industry output of more than 20 million vehicles, but total cars sold in the United States in 2002 and 2003 numbered fewer than 17 million—2004 sales are forecast no higher.[8] There are some 32 million more automobiles registered in the United States than there are licensed drivers.[9] The notion that General Motors (GM) spent $3.43 billion on advertising in 2003—the largest ad spend in the world—simply illustrates how hard corporations must work to uncover sources of demand.[10] Procter & Gamble (P&G) may be the world's leading consumer packaged goods company, but it spent more than $3.32 billion on advertising in 2003, largely just to maintain market share. This meant that P&G spent $1 on advertising for every $13 of revenue.[11] The simple truth is this: The scarce resource is no longer supply, it's demand. With so many companies vying for a finite number of customers who have limited time and attention, you might conclude that where supply was finite and demand was infinite in the 1950s, supply is infinite and demand is finite today.

Disappearing margins

We know that most companies today are struggling to maintain margins. Whether in the hypercompetitive PC market or in automobiles, the pressures are similar. Consider new-car rebates. In July 2004, the average rebate package among global brands was over $3,000 per vehicle versus a little more than $1,500 in August 2000; for U.S. automakers alone, the average was above $4,000 per vehicle.[12] As far back as the late 1980s, this was already an issue at GE Medical Systems. Then, as now, this unit of General Electric (GE) was one of a handful of globally dominant makers of medical-imaging machines, which included MRI machines and CAT scanners. Despite the pathbreaking and lifesaving nature of its equipment, the economics of the business were sobering. If GE sold a million-dollar machine outright, it realized no margin; if GE sold a machine bundled with services, it made money. GE's approach was and is to sell every machine it can with a financing plan, an extended warranty, a maintenance plan, and an ongoing program for training and certification. Without a service-based relationship, the business loses money—even in a global oligopoly making equipment that literally works miracles and saves lives. Many industrial behemoths are heeding the logic here, which implies that margins exist not in what we sell but how we sell it—in the ongoing relationships wrapped around the offerings. GE's Jeffrey Immelt has committed his company to generate 50 percent of revenues from services by 2005.[13] IBM under Lou Gerstner and now Sam Palmisano has shifted the company's activities from sales of "big iron" to a service-oriented model based on computing on demand, which has fueled the growth of its massive IT services unit, IBM Global Services.[14] A similar emphasis on relationships over offerings is why chemical companies now sell cleaning solutions instead of chemicals to customers. And it's why William Clay Ford, Ford Motor Co.'s scion and CEO, has defined the business of his family company as providing transportation solutions instead of building cars.

The net effect of these three trends—accelerating product and service life cycles, supply outstripping demand, and disappearing margins—highlights a key insight: In a world of offering-based advantage, the quality of a company's interactions and relationships with its customers is important but not often mission-critical. To coin a phrase, when a company truly does sell a better mousetrap, the world—given adequate exposure through advertising and promotion—will beat a path to its door. But when a company operates in a market dominated by world-class competitors selling world-class mousetraps, the interface system can represent the only truly viable and defensible source of competitive advantage.

DRIVERS OF TECHNOLOGY AS RELATIONSHIP MANAGER

In posing the obvious follow-up question—how might companies manage their interface systems for maximum competitive advantage?—we turn to the changing landscape of technology. This brings us to the threshold effect created by four trends affecting interface technologies, which together enable *interface capability* as the new basis of competitive advantage. It's imperative to understand these trends because they represent the basis of the opportunity we've discussed—to enhance the quality of interactions and relationships between companies and their customers while simultaneously driving down the costs of managing those interactions. The changing attributes of new interface technologies mark a change in the efficiency and effectiveness frontiers at which corporations can manage interactions and relationships with their customers and markets. Elaborating upon our earlier discussion of networks and devices, these four technology trends are:

1. Proliferating smart devices

2. Rising intelligence and interactivity of such devices

3. Increasingly affective appeal of devices

4. Near ubiquitous global connectivity of information networks that serve to connect such devices

None of these trends, in and of itself, is new. Each of these aspects of technology evolution has been unfolding for years or even decades. Together, however, they represent a more detailed and nuanced picture of the dynamics that we touched upon in chapter 1. What's noteworthy here is the combined threshold effect: The opportunities it creates for businesses are not only new but unprecedented—because they will reconfigure the very nature of work.

Device Proliferation

As nearly everyone who's not living an isolated Thoreauvian existence knows, semiconductor-based processing power continues to increase dramatically as a function of price and size. With all-too-familiar logic, Moore's Law observed that every eighteen to twenty-four months the semiconductor-manufacturing industry could double microchips' processing power in relation to price. This made computing power more susceptible to integration in an ever widening variety of consumer and business devices. Such increased power in smaller packages has resulted in a plain reality: Microprocessor-based or microprocessor-enhanced devices are everywhere—and their prevalence is rising among consumers and businesspeople alike throughout the industrialized world. As *The Economist* has commented, "Automated machines have . . . quietly slipped into many corners of everyday life."[15] The dynamics of Moore's Law have effectively delivered a reality that has defied sober prediction, such that most computing power in the world around us is no longer visible—unlike those fabulously impressive computers and robots in science fiction of yesteryear—because it's embedded throughout our world. As *The Economist* concluded:

> The idea that [computers] would become cheap enough to be integrated into almost any specialized device, from a coffee-maker to a dishwasher, was hard to imagine. Instead, it seemed more likely that such intelligence would be built into a small number of machines capable of turning their robotic hands to a range of different tasks. In place of the general-purpose housebot, however, we are surrounded by dozens of tiny robots that do specific things very well. There is no need to wait for the rise of the robots. The machines, it seems, are already among us.[16]

Enter any meeting, mall, airport, or home, and you can see this dynamic at work. Businesspeople carry mobile phones to make calls, BlackBerrys to check e-mail, Palm devices to check calendars, not to mention an array of Pocket PC–based devices from Hewlett-Packard, Microsoft, Sony, and Toshiba to perform similar functions. Consumers carry MP3 players such as iPods on their belts, mobile phones that snap and send digital images and video clips, and high-tech watches that double as cameras, calculators, and MP3 players. Laptops, too, have become so small and powerful that, even as they supersede the functionality of high-end, desktop-style workstations, they are portable, personal, and even stylish to carry. Intelligent devices that can sense and respond, "think" and communicate, are everywhere—and this state of affairs is only becoming more pronounced. In Japan, the NTT DoCoMo i-mode handsets, mobile phones with wireless access to digital downloads and online service, have diffused more rapidly in their home market than any product in history. Before 1999, the i-mode did not exist. As of June 2004, more than 41 million Japanese consumers used the device to message one another by voice and text, to entertain themselves, and to otherwise manage their lives.[17] While i-mode's adoption in Japan is impressive, it's even more stunning to note that China has become the world's largest mobile phone market—with nearly 270 million users in 2003 or an installed base roughly equivalent to the population of the United States.[18] Worldwide adoption of mobile phones is expected to top 1.5 billion in 2004—that's equivalent to one-fifth of the world's population. Similarly stunning figures estimate that more than half a billion PCs and hundreds of millions more specialized computers (otherwise known as game consoles from Sony PlayStation, Microsoft, and Nintendo) operate around the world (see figure 2-1).

In short, the presence of smart devices in our lives is no longer the exception but the rule. And the research figures we cite take into account only those devices that exist explicitly to deliver processing power. These figures do not include the devices that rely on microprocessors to accomplish physical tasks, like household appliances and office machines; or packaged goods that contain electronic or RFID tags; or smart cards that store value or users' identities. Device proliferation is not about any one technology or medium, but about the presence of intelligent technology—from microprocessor-based devices to computers embedded in other goods—everywhere.[19] This proliferation points to the next trend: The devices that surround us are more intelligent and interactive than ever before.

FIGURE 2-1

Worldwide Adoption of Devices and Networks: 1991–2003

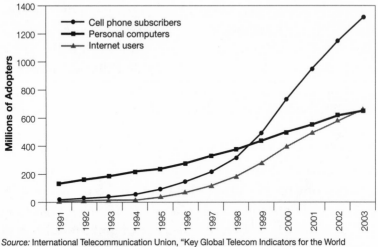

Source: International Telecommunication Union, "Key Global Telecom Indicators for the World Telecommunication Service Sector," International Telecommunication Web site (<http://www.itu.int/ITU-D/ict/statistics/at_glance/KeyTelecom99.html>).

Intelligence and Interactivity

Although Moore's Law cannot hold out indefinitely, Intel has indicated that it is still on track to double processing power every two years through about 2011. Continued development of other transistor insulators, such as strontium titanate, can push this date out even further.[20] Intelligence, of course, enables devices to process data as it is received and to perform specific tasks with that data; interactivity enables smart devices to operate in a sense-and-respond mode. Such intelligence and interactivity create the potential for devices to perform complex tasks, including, among other things, interacting with and relating to human users. Consider what's happening in the field of robotics, where commercialization is finally reaching into the consumer realm.

Robots are that quintessential convergence product that arguably brings processing power, its intelligence and interactivity, together with service interfaces. After all, robots are, in their original definition, machines that do things to serve people. Companies such as Japan's Fanuc and Kawasaki have been making sophisticated industrial robots for manufacturing automation since the 1980s. Estimates suggest that approximately one million industrial robots were in operation around the world in 2002, with nearly half of them in Japan, 233,000 in the European Union, and 104,000 in the United States.[21] Fanuc and Kawasaki, however, have never designed robots to interact with consumers, only with production lines and workers. Moreover, their products take form mostly as mechanical arms; they are devices with limited intelligence, circumscribed interactivity, and no anthropomorphic interface attributes, such as speech, humor, or personality—even if they can look surprisingly alive on plant floors.

More recently, a new wave of technology companies has entered the fray with consumer-oriented applications of robotics, resulting in 545,000 entertainment or leisure robots by year-end 2002, out of a total of 625,440 service robots targeting consumer applications.[22] Entertainment robots, like Sony's AIBO robotic dog, are the fastest-growing and largest consumer category, with projections of 1.5 million sold by 2006. In the same time period, projections show consumers adding to their homes 400,000 domestic robots that vacuum and 125,000 that mow lawns.[23] We've already referenced Roomba, the saucer-shaped vacuuming robot, which wanders through consumers' homes in charming and unobtrusive ways—a phenomenon that some owners find captivating (especially at a retail price of just $200). Wakamaru, a household robot from Mitsubishi Heavy Industries designed to provide companionship for older people, recognizes individual faces and engages in dialogue with users.[24] Banryu, which translates from the Japanese as "guard dragon," is a domestic robot developed by Sanyo Electric Co. in a joint venture with tmsuk Co. to protect consumer homes from intruders, gas leaks, and fires.[25] All of these products are recent arrivals to the market. Earlier consumer-friendly robots appeared as denizens of retail environments. For instance, Schrobbie, a dark blue cartlike robot, has been cleaning and drying floors for the Dutch retail chain Albert Heijn since 1996—while the stores are open for business. If a shopper stands in its way, the Vespa-like vehicle will say, "Excuse me; I'd like to clean the floor here." If the customer fails to move, Schrobbie will navigate around her.[26] Another early example is a cartlike robot (it resembled a hot dog stand that had escaped from mid-town Manhattan without its owner) that used to serve drinks in the stylish London-based chain of restaurants called Yo! Sushi. At one flagship store, it toured the periphery of the bar carrying a variety of chilled beverages, using visual sensors to follow a yellow line on the floor. If you stood in its way, it would stop and ask you, in a science-fiction-circa-1950 robotic voice, to select a beverage from a collection atop its stainless steel carapace. If you stood in its way for too long, it would politely quip, "Please move. You are standing in the way of the future." It was hardly impressive by today's standards, but the cart had enough personality to enthrall patrons, while getting the job done.

None of these robots, however, has the mass-market appeal of AIBO, developed by Sony as the world's "first entertainment robot." AIBO is a robotic pet; its name refers to an AI-enhanced, or artificial intelligence, (ro)BO(t). Introduced in 1999, it looks like a small mechanical dog that moves on all fours, hears voice commands and responds appropriately, sees walls and avoids furniture, and learns from its master's interactions to yield, over time, a unique personality. Hardly larger than a football, the original AIBO had more processing power than most desktop PCs today. The earliest model came equipped with a 64-bit RISC processor, 16 megabytes of internal memory, 8 megabytes of additional memory in a removable format, and eyes enabled by a 180,000-pixel color video camera and infrared sensors that could estimate distances and recognize shapes.[27] Even today, it's one of the most powerful PCs you can buy, even if it does walk on four legs. The latest versions can dance to music, remember faces and voices, sing on remote command, and send pictures through its eyes over Wi-Fi connection to owners.

Of course, this kind of intelligence and interactivity is hardly confined to the domestic-robot market. Such processing power is all around us. Consider, for example, the BMW 745i, introduced in 2002. This top-of-the-line luxury sedan had so many microprocessor-controlled, software-defined attributes that it nearly qualified as an AIBO on wheels. The owner could customize the car's performance and features by accessing more than seven hundred functions from three hundred pull-down menus on the largely button-free dashboard, which was dominated by a centrally located flat-panel display.[28] This system, known as iDrive, was the pinnacle to date of computer-enhanced automotive engineering, and its integration with the car's physical performance attributes reached new heights of bundled hardware-software design. Of course, such complex systems are not without their problems. This is what BMW learned from affluent U.S. consumers who adopted the 745i; for many, the car was too complex, requiring dealers to run seminars before new owners could successfully drive off the lot. When Hollywood celebrities got their hands on this sought-after vehicle, they naturally drove it someplace they could valet-park it. Unfortunately, L.A.'s parking attendants could not manage to move the car from curbside to parking lot, because they could not figure out how to drive it. BMW politely met this need by providing 745i owners laminated wallet cards for valets that displayed step-by-step instructions for basic vehicle operation.

Devices characterized by high levels of intelligence and interactivity have arrived—and some are beginning to respond, interact, and adapt to the needs of their human owners as if they were actual life-forms. All that's missing from the equation is that fundamental attribute of humanity—emotion.

Affective Appeal

Products such as AIBO are notable for the relationships they can establish with users based on the emotional responses they elicit. For example, AIBO has become the organizing force behind a social phenomenon. The product has given rise to new forms of community affiliation among its owners, such as AIBO robot clinics, AIBO birthday parties, and AIBO soccer matches.[29] In popular destination cities, owners—with their AIBOs—meet and spend time with one another at conventions. What's strange is not the fact of the convention itself. After all, *Star Trek* fans attend Trekkie conventions, and Harley-Davidson motorcycle riders join Harley Owners Groups. What's noteworthy is that, unlike Harleys, which owners customize and collect over the years, AIBOs pretty much look, and are, all alike. (There are multiple colors of AIBO, and newer ones have more advanced features, but they are all essentially the same machine.) The AIBO, however, is "born" with the basic personality of a puppy. When it comes out of its box and owners turn it on for the first time, it just sits for several hours and does nothing. Remarkably, each unit is endowed with sixteen thousand latent personality attributes, which are activated, or not, through interaction with its owner. Given the serendipity of such interactions, each AIBO develops an essentially distinctive personality a few weeks into its life, thus becoming unpredictably and unaccountably unique, like a real pet. As a result, the conventions enable owners to see their AIBOs interact with one another and shape one another's already distinctive personalities, while owners presumably enjoy interacting with each other. The impact in emotional terms is arresting: Many owners reportedly develop a conviction that AIBO is alive. They name their AIBOs and refer to them not as "it" but as he or she. They talk about them not as "robots" or "toys" but as dogs or pets. Many experience a heightened sense of emotional attachment that normally would not accrue to, say, a microwave oven or a digital camera.

Explicit experimentation with the potential for machines to arouse seemingly authentic emotional responses in humans has grown sophisticated in recent years. Not all interactive toys attempting to appeal to the hearts and minds of children succeed. For example, the much-touted Interactive Barney was a flop. But researchers persist. A case in point is a creature called Affective Tigger at the MIT Media Lab. Using AI-enhanced computer reasoning in analyzing physical cues, Tigger can interpret a child's mood, recognizing what it feels like when a happy or a sad child picks him up and responding appropriately through sounds. Such socially perceptive reasoning has reached greater extremes in the lab's development of Kismet, devised to push the limits of affective computing. Kismet is a physical interface that resembles a human face capable of rich expression; it can read a user's facial expressions through visual sensors, listen to a user's words, and respond through facial expressions of its own in socially appropriate and emotionally authentic ways. If Kismet sounds like an impractical high-tech toy, consider the related technologies already at work in some automated call center systems. Researchers such as Shrikanth Narayanan, professor in the Speech Analysis and Interpretation Laboratory at the University of Southern California, are developing software programs that monitor the emotional tone of callers as they interact by phone with speech recognition-based services; when the software recognizes words or tones of voice that suggest anger or frustration, they trigger a default to a human operator, who can then deal person-to-person with the upset customer.[30]

The experience of interacting with such devices can result in the kind of "willing suspension of disbelief," to cite the words of Coleridge, that literature invites—except that it can often be *un*-willing. That's what AIBO, Affective Tigger, and Kismet do: They blur the line between what we feel, willingly and naturally, for living things—a connection that is powerfully emotional, even primordial—with what we feel for inanimate objects. It's the threshold beyond intelligence and interactivity that smart devices have begun to broach—the threshold of affective attachment and emotional response that normally belongs to living things. Machines appeal now to our hearts *and* our minds. And this is what's driving some leading corporations to experiment with high-end robots, priced like luxury cars, such as Honda's ASIMO (which stands for advanced step in innovative mobility as well as a homage to science fiction writer Isaac Asimov, author of *I, Robot*) and Sony's Dream Robot (formerly the SDR-4X, now called the QRIO). Both are mobile bipeds capable of navigating the physical world of human activity in uncannily human ways and relating to humans, from whom they learn over time.[31] The QRIO is particularly advanced in features and functions. It has both short-term and long-term memory, which enables varieties of learning, and it can recognize up to ten human faces in complex visual settings, identify humans by tone of voice, and, using seven microphones embedded inside its head, locate the direction or source of a voice or sound. It can also use its wireless LAN connection linked to software running on a PC to operate continuous speech recognition in multiple languages. And it can run as well as walk.[32]

Neither of these products is in any sense ready for consumer mass markets. But both signal the power of automation and machine intelligence when combined with seemingly vital attributes, such as learning, mobility, and human or animal form.[33] Honda deploys ASIMO for ceremonial functions, for example, ringing the opening bell at the New York Stock Exchange in February 2002 for the company's twentieth-fifth anniversary on the exchange.[34] On a recent visit to a state dinner in the Czech Republic, ASIMO accompanied Prime Minister Junichiro Koizumi to greet Czech dignitaries with this comment: "I have arrived in the Czech Republic, where the word robot was born, together with Prime Minister Koizumi as a Japanese envoy of goodwill."[35] At home at Honda's Tokyo headquarters, ASIMO also greets VIP guests and leads them to conference rooms for meetings. The only problem, Honda reports, is that people will often follow ASIMO out of the conference rooms after they have arrived to see what else he will do.[36]

Of course, the affective appeal of certain physical products is not new. As consumers, we will often describe ourselves as emotionally attached to things because they look lifelike, friendly, or engaging, even if they do nothing (Steiff bears); because they are sleek, intuitively designed, and easy to use (Apple's Macintosh); because they are ergonomically appealing in surprising ways (OXO Good Grips); because they augment an image of ourselves (the Hummer H2); or because they appeal strongly to our aesthetic sensibilities (the Palm V). Nearly a decade ago, Jeff Hawkins, the cofounding chief technologist of Palm Computing, conceived of a personal digital assistant (PDA) that would be every bit as functional as previous handhelds from Palm but would be smaller, slimmer, and *beautiful*. Thus was born the Palm V, which energized the category selling 6 million units from 1999 through 2001, when Palm discontinued it.[37] Users responded to the V's beauty by finding ways to integrate its elegance into their lives. That created a market for luxury goods makers Louis Vuitton and Coach to design leather fashion cases for the Palm V, fusing their brands with Palm's. When things like that happen, technology crosses a line; it transcends the personal and becomes intimate.

It's clear that objects that stimulate strong aesthetic, ergonomic, or emotional responses have a unique purchase on our time and attention. When those responses converge, in turn with intelligence and interactivity, there is a potential for even more complex bonds to be formed. A child loves a plush animal not only because it's adorable; she imagines it has a life of its own and conjures up an endless variety of make-believe interactions. An adult who uses a compelling digital device becomes emotionally attached to it not only because it's pleasurable or attractive to use; he interacts with it all the time, processing—to him—mission-critical information. (How often have you heard someone say about her PDA, laptop, or smart phone, "My *life* is in that thing!"?) Even a relationship with an automobile is shaped by the fundamental interactivity involved in driving; we do things to the car, and the car does things to us. When the aesthetic or emotional dimension is reinforced by intelligence and interactivity, it's clear that these dynamics have important implications for companies aiming to create, manage, and augment truly meaningful relationships with customers. Devices that combine emotional appeal with aesthetic or ergonomic satisfaction—especially if they are also sufficiently intimate that we can wear them—are particularly powerful. What's left is to connect those devices to one another to render them even more relevant and useful—that's where networks come in.

Ubiquitous Connectivity

Much ink has been spilled in recent years on the emergence of the Internet as a new economic and social infrastructure—as great in its impact as any of the grids that came before it, such as power, transportation, and telephony. And it came of age with enormous speed. It took only three years to reach 50 million U.S. homes, as compared with radio, which took thirty-eight years, and television, which took thirteen.[38] Additional comment is hardly merited here. It's important, however, to observe that the proliferation of digital devices alone, even with their rising capabilities of intelligence, interactivity, and affective appeal, can take us only so far into the realm of relationship management without the ability to connect them to what is, in effect, a digital nervous system such as the Internet. Broadband allows us to serve more information more quickly to such devices or interfaces, and we have only to look to leading markets in Japan and South Korea to see that changes in degree—in this case, access speed—can result in changes of kind, as realized through user experience. (In those countries, broadband penetration leads the world, and broadband speeds are 5 to 20 times faster than in the United States.) Bluetooth allows owners to create mini-networks with devices in their homes and offices. Various versions of Wi-Fi, and now WiMax, increasingly enable us to connect any smart device to any network anywhere—local or global.

Connectivity takes various forms, including people to machines, machines to machines, and people to people through machines. For example, in South Korea, high-speed broadband connectivity in about 70 percent of homes has given rise to a social movement of massive multiplayer electronic games, rendered in rich-media interactive environments, making Seoul the world's center of excellence in software development for the next generation of game platforms.[39] Games are a category where intelligent, interactive, and affectively appealing characters, environments, and competition—which already had developed a following of enormous scale based on stand-alone console games—are undergoing a renaissance spurred by ubiquitous connectivity. Sony's PlayStation 2, Microsoft's Xbox, and Nintendo's GameCube have each stimulated new waves of demand with offerings that link users over networks to one another. In April 2003, 21 percent of players using the PlayStation 2 game *NFL 2K3* played online, and data from the Entertainment Software Association indicates that 43 percent of avid U.S. gamers were playing online in 2004.[40] The appeal of such games—wherein users interact not only with other users, but also with, and as, vividly realized characters on-screen— is testament to the might of processing power and affect combined with connectivity. These machines enable users to share entertainment environments in which they spend hundreds of hours engaged with games and their characters. (These worlds occasionally give rise to character franchises that are powerful enough to jump industries, such as *Tomb Raider*'s Lara Croft, who went from games to motion pictures and became, however briefly, a hot property in Hollywood.)

The social phenomenon of online gaming is a juggernaut that has made electronic games an industry larger than Hollywood. (Global video games sales hit $35 billion in 2003, while the world-wide box office gross was a little less than $15 billion.)[41] Nonetheless, gaming is but one example of connectivity's power, over and above the evolution of smart devices themselves. The penetration of residential broadband connections has momentum to drive all manner of innovations, with compound annual growth rates from 2000 to 2004 of 109 percent in Europe, 101 percent in Latin America, 77 percent in Asia, and 50 percent in North America (see figure 2-2).[42] Whether it's the fact of online penetration reaching more than half of homes in many industrialized nations, or the fact of mobile phones reaching similar thresholds among adult users in most major country markets, the linkages among smart devices have brought the information resources of global networks into our physical experience of everyday life.

If the dot-com era was fueled by an appreciation for the power of networks online, today we are concerned with the power of

FIGURE 2-2

Broadband Proliferation

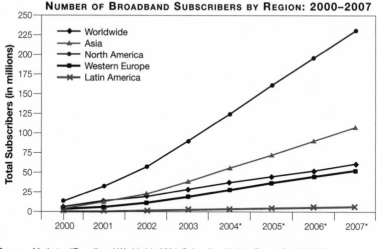

NUMBER OF BROADBAND SUBSCRIBERS BY REGION: 2000–2007

Source: eMarketer, "Broadband Worldwide 2004: Subscriber Update Report," 4 April 2004.

* Indicates estimate

networks in our lives *everywhere*. That's why short-messaging service (SMS) has changed the way Europeans walk down city streets, half-looking ahead and half-gazing into the tiny screens of their mobile phones. It's why online youth around the world—and increasing numbers of corporate users—are in real-time communication with multiple friends and coworkers simultaneously using instant messaging on PCs and handheld devices. It's why retailers such as Banana Republic and Old Navy can enhance the productivity of clerks by requiring them to wear wireless headsets as they move through the stores, interacting with customers while consulting one another and fielding incoming phone calls. And it's why we can begin to talk about home automation in sophisticated forms, where audio entertainment can follow inhabitants around their homes, where lights and window coverings self-adjust based on external weather conditions and time of day, and where household appliances can report to central servers on developing problems or maintenance issues.[43]

The phenomenon of ubiquitous connectivity—which links intelligent, interactive, and affectively appealing devices—means that such smart devices need not manage relationships with customers on a stand-alone basis. As interfaces designed for relationship management, these devices can maintain ongoing connections between customers and the companies and brands that spawn them. Consider GM's OnStar System. OnStar allows drivers to speak to GM no matter where they are at the touch of a button—and for GM to speak to them. That's a radical innovation, and one that acknowledges that for the owner of a new Cadillac, the car—not the TV advertising or the billboards or the dealers—is the primary customer relationship management interface postpurchase. It's how the car interacts with the customer—and how much of an emotional bond it's able to form with its driver—that will largely determine the likelihood of repurchase. Similarly, consider TiVo, the personal video recorder. TiVo maintains periodic connectivity with its corporate servers, reporting consumer usage data, updating billing information, and generating programming recommendations. It is another interactive relationship manager that acts as a gateway, in real time, between the company and its customers. Consider also Windows XP. It provides an interactive Help function that's only available online; as a result, every user who seeks assistance with the latest operating system must interact directly with Microsoft itself. The product is the relationship manager, until there's a problem; then XP becomes a gateway for dialogue with the corporation that supplied it. Of course, each of these examples is also an illustration of how ubiquitous networks create the capacity for people and machines to serve customers from remote rather than proximal locations, which is what's making possible the outsourcing of frontline service jobs to foreign shores and frontline service functions to distant data centers.

The critical insight here is that two streams of technology—devices and networks—have converged. Connectivity has become increasingly ubiquitous through waves of innovation starting with the invention of the telegraph in 1837, the telephone in 1876, broadcast media in the early twentieth century, and the Internet in 1969. Similarly, smart devices have steadily gained processing capacity ever since the invention of the transistor at Bell Labs in 1948. Combined, these two occasion the truly remarkable: Networks and devices have done more than create an online world. Together, they have permeated our offline lives, and their capabilities are sufficiently smart, interactive, and appealing; and their connectivity is sufficiently flexible and diverse (wireline and wireless, local and global) and comprehensive (users to users, users to machines, machines to machines). The result is that we have achieved a kind of threshold effect or tipping point. The new possibilities for devices and networks have expanded sufficiently that a difference of degree has become a difference of kind. In this sense, these developments in technology irrevocably alter the possibilities for how companies interact with their customers on their front lines. By enabling a work force of people and machines in proximal and remote locations, these trends create the basis for the radical reconfiguration of work for companies—and for the vast majority of the people they employ.[44]

CONCLUSION

What a company sells—its offering—may be susceptible to commoditization, but the interactions and relationships companies establish with their customers are not. Interactions and relationships are the context in which otherwise commoditized, or rapidly commoditizing offerings, are differentiated from companies. Unless what you sell defines your advantage in sustainable ways, service interfaces matter—now more than ever—in establishing and maintaining competitive advantage. That's why we call this chapter the *interface imperative*. And that's why we focus on the potential for reconfiguring how companies may deploy the interfaces that will define their market positions in the future.

- *Because microchips have risen in processing power, fallen in cost, and achieved new thresholds of miniaturization, smart technology is all around us.* It's not a matter of waiting for technology to catch up with the vision articulated in this chapter. Ubiquitous embedded intelligence is a reality. And that has resulted in a wide variety of products—from hi-tech toys to consumer durables—that are capable of interacting with customers in meaningful ways as never before.

- *Companies must concern themselves now more than ever with management of customer interactions and relationships because they face threats associated with an era of total commoditization.* Product lifecycles have accelerated in nearly every industry sector, making differentiated offerings generic or obsolete faster than ever before. Sector after sector in the economy suffers from overcapacity. And margins, even in highly sophisticated technology products, are difficult to maintain. That means companies must establish advantage—in market position, pricing power, and brand—in new ways.

- *Reconfiguring a company's interfaces with its customers has risen in importance just as technology trends related to devices and networks have achieved a threshold effect.* There are four trends we see that have contributed to the potential for radical innovation in how companies relate to customers and markets. These are the proliferation of smart devices; their rising intelligence and interactivity; their increasingly affective appeal; and the ubiquitous connectivity that connects those devices, wired or wirelessly, to intelligent networks and to one another. Taken together, these trends spell potential gains in the efficiency and effectiveness with which companies can manage interactions and relationships with customers. For this reason, technology is driving a relationship revolution that's also a productivity revolution. That's why reengineering—not of the back office but of the front office—is the focus of chapter 3.

3

THE FRONT-OFFICE
REVOLUTION

T HE FRONT-OFFICE REVOLUTION is an industrial revolution. It affects the vast majority of the work force, and it involves a vast array of work-altering new technologies. It drives companies to rethink the fundamental economics of their businesses. Adapting to this kind of change, in any period of history, has proven daunting. Indeed, business is rife with scenarios illustrating how dislocating periods of change in the nature of work can be. Consider the British cotton textile industry in the late nineteenth century. Under threat from American mills in New England that were using stream-driven automation to produce cotton textiles, the British simply failed to adapt. Preserving the old ways of doing business proved more compelling than the harsh economic logic of industrial automation. In less than a decade, the British textile industry collapsed under the assault of New England's more efficient producers.[1] In many such capital-for-labor substitutions, social reaction has been fierce: In Britain, the Luddites famously organized workers against the newly mechanized modes of production—and created a new word in the English language for those who resist change.[2] In the United States, the Homestead Strike in1892 targeted the hard driving practices at the Carnegie steel mills, and the General Motors sit-down strike in the 1930s protested labor relations at the automotive assembly plants. Even the information industries have had their share of such labor uprising. Recall the New York Times Company in the 1960s, which faced a walkout of the newspaper's linotype operators who were protesting the transition to photo-offset production processes, soon to become the industry standard for newspaper and magazine production around the world.

We are experiencing analogous labor strife today—except that this time the jobs in question are in the front office. For example, the strike and lockout involving the International Longshore and Warehouse Union, which immobilized twenty-nine West Coast ports in the United States, targeted the preservation of what were, in effect, white-collar jobs. The last deadlock had occurred in 1971, when the union attempted to block the introduction of container cranes that would ultimately reduce the West Coast work force from more than a hundred thousand workers to just ten thousand five hundred today.[3] In mid-2002, the union opposed automation that would allow computer operators to control movements of cargo on and off ships from remote locations miles away from any one port, speeding the cargo handling. The union called for a work slowdown, and the Pacific Maritime Association, a group of port operators and commercial shipping lines, retaliated by locking out employees. Ultimately, the Bush administration invoked the Taft-Hartley Act, which allows presidents to seek injunctions blocking strikes that "imperil the national health or safety."[4]

Earlier industrial revolutions affected huge swaths of economic activity, involving large segments of the work force in both agriculture and manufacturing. Today's revolution, however, arguably affects the largest job category ever altered by technological change: services. The threshold effect involving devices and networks allows companies to displace work, automate it, or both. The front office is where services take place. The opportunity to reconfigure it is real, but not without cost. That's why any mention of off-shoring jobs grabs headlines; most of the labor force, white collar and blue collar, work for a living in the front office. Consider recent news headlines regarding such labor displacement: Gone offshore are twenty-four-hour call centers and help desk positions (American Express, Bank of America, Delta Air Lines, Dell, Kodak); analyst positions at investment banks (Bear Stearns, Lehman Brothers); research and development jobs (Boeing, GE Medical Services); tax preparation (Procter & Gamble); software development (Microsoft); chip design (Intel, Texas Instruments); medical diagnosis of radiological scans (Massachusetts General Hospital); and architectural design (Fluor Corp.).[5] One research firm projects that some 3.3 million white-collar jobs and $136 billion in payroll will shift from the United States to lower-cost labor markets by 2015.[6] For one call center in India, operated by Spectramind (a unit of the giant Indian systems integrator Wipro), a highly skilled customer service representative with a college degree earns an annual income of $3,710, nearly an order of magnitude less than the going rate in the United States. Bank of America can perform work that costs $100 an hour in the United States for $20 an hour in India.[7] By early 2003, Spectramind employed four thousand people, many of whom received voice and accent training to work effectively with North American consumers, to handle businesses that included processing insurance claims, selling collectibles by phone, and reviewing scientific research for pharmaceutical companies.

While we can attribute such reorganization to ubiquitous networking, there is another force of perhaps equal or greater impact, associated with the combination of global networking and smart device proliferation: front-office automation. When robotics, for example, threaten to alter work processes behind pharmacy counters, people notice because white-collar service jobs are in peril and because pharmacy counters are often intimate service settings.[8] This *front-office revolution*—a revolution in how businesses harness front-office technologies to manage customer interactions and relationships—holds equal potential for labor strife. For example, the New York City Transit Authority announced plans to close 177 token booths in the city's subway systems and replace them with MetroCard vending machines as well as automated turnstiles in 2003. Such automation would eliminate 450 jobs and save the system, which was suffering huge deficits, $25 million a year. Opposition came not only from the Transport Workers Union Local 100 but also from community and transportation advocacy groups. (In the end, 62 booths closed in 2003, but another 164 will be closed by 2005.[9]) Despite resistance, however, such changes are inevitable because the focus is not on efficiency alone. The rewiring of front-office work across every sector of business—through displacement and automation—is as much about effectiveness (the quality of customer interactions with service providers) as it is about efficiency (as measured in lower operating costs).

The front-office revolution represents a new way of conceiving of customer relationship management (CRM). In business lingo *CRM* is a technical term, referring to enterprise software systems designed to manage a variety of front-office customer interaction functions. In our view, that definition of CRM is too limited. CRM should describe everything that people and machines working together in an organization do to interact successfully with customers—thereby establishing meaningful relationships. Of course, third-party software and systems may prove valuable elements of a company's customer relationship management activities, but CRM, writ large, is not something a company can or should outsource. It's a critical driver of competitive advantage. Thus, CRM is what entire companies do to compete—deploying interfaces in optimal ways involving people and machines configured according to new possibilities for displacement and automation. By this definition, CRM should become the rallying cry for corporations; the interfaces they deploy with customers will ultimately determine most companies' competitiveness over time. Those interfaces can achieve efficiency and effectiveness outcomes when deployed as optimized interface systems, which are integrated according to a company's goals with respect to interactions and relationships with customers. Of course, the question of how those front offices operate is both more critical and more complex than ever before.

The implications of the front-office revolution are far reaching, as they inexorably alter the enterprise economics of companies; the approach to strategy, organization, and culture; and, ultimately, growth. This revolution, like its predecessors, is about a great leap forward in productivity, and it begins with the task of reengineering the front office.

In the balance of this chapter, we'll examine what it means for companies to reengineer their front offices, based on the underlying logic of displacement and automation that networks and devices enable, and how senior managers may address these opportunities in strategic and operational terms.

PRODUCTIVITY GAINS FROM FRONT-OFFICE AUTOMATION

One might argue conceptually that productivity is composed of two elements that determine companies' ultimate ability to deliver value to customers. These are performance, defined as results achieved for customers, and cost, defined as the resources companies must expend to deliver performance and which ultimately influences price. Some have referred to this ratio of performance over price as a customer "value equation."[10] Alternatively, we would characterize this equation as a ratio of effectiveness over efficiency. Radical productivity gains often translate into radical enhancements in customer value propositions, and they usually require dramatic movements in the numerator (up) and the denominator (down). Front-office reengineering aims to do that. Leveraging networks and devices to reconfigure front-office or service-related tasks, reengineering can potentially change both the quality of customer outcomes and the costs of delivering them. Earlier, we considered several companies, including FedEx, Progressive Insurance, and Krispy Kreme, that had realized gains primarily in efficiency by reconfiguring their front offices. Now, consider the experience of several corporations that have used front-office automation technology primarily to drive effectiveness—and top-line growth:

- *McDonald's* is experimenting with self-service kiosks in some restaurants, enabling its customers to use touch screens to manage their own order entries. While kiosks may drive down the costs of taking orders, they have proven surprisingly robust as a strategy to satisfy customers. Patrons appreciate the kiosks for "the opportunity to use McDonald's the way they want to," as they enhance convenience, accuracy, and control. According to McDonald's, these interfaces enable it to process orders more quickly and cross-sell or up-sell more effectively. The results: McDonald's reports an average self-service order is $1.20 higher than the average order placed with its employees.[11]

- *Borders* has deployed self-service kiosks in its book superstores enabling customers to search both in-store and catalog inventory. Dubbed Title Sleuth, the three hundred kiosks, which the chain introduced in 1998, are distributed across most of Borders's 360 chain stores and were serving up 1.5 million customer searches a week by 2004. Nearly one of every three customers in the stores uses them, providing access to 15 million titles (the stores carry only 200,000 titles). The results: Customers who use Title Sleuth, which Borders believes contributes to its brand consistency and differentiation, spend 50 percent more during each store visit, and they generate 20 percent more special-order sales.[12]

- *Recreational Equipment, Inc., (REI)* began placing kiosks in its sixty-five stores starting in 1997; its newest stores have four kiosks each. A typical REI retail operation has thirty thousand SKUs on display, but the company sells seventy-eight thousand SKUs on its Web site and in its catalog. The kiosks provide better information at its points of sale, which REI believes translates into better in-store service for its customers. As one manager puts it, "No matter how smart [our store clerks] are, they can't keep 45,000 pages of information [in their heads]." Though REI has done nothing to promote the kiosks, it offers free delivery on Web orders if customers pick them up in its stores. The results: Kiosk-mediated sales at REI are growing 30 percent a year, building revenues to date to the equivalent of an additional 25,000-square-foot brick-and-mortar store.[13]

- *Kroger* has turned to self-checkout systems to better serve customers in its grocery stores. From a cost perspective, its U-Scan systems deliver immediate labor savings, enabling one employee to oversee four checkout lanes in place of four cashiers. The U-Scan machines have also helped Kroger create an innovative corporate image. The results: At $150,000 for a four-scanner system, Kroger has realized rapid breakevens on the equipment based on labor savings, but the real gains derive from increased revenues, based on speed of service and enhancement of Kroger's brand image.[14]

- *Rite Aid* is experimenting at pharmacy counters with robots dispensing prescriptions and voice-response units (VRUs) to process prescription orders from patients. The drugstore chain has turned to automation in the face of an anticipated labor shortage: Prescriptions filled in the United States will grow by 30 percent over the next two years, while the number of pharmacists will likely increase by only 6 percent. Using robotic dispensing systems, a chain pharmacy can raise its production capacity from two hundred to seven hundred prescriptions a day and call on pharmacists to perform less menial labor behind the counters. The results: The combination of front-office automation techniques—VRUs, Web sites, auto-fax systems, and robotics to process patient orders and deal with physicians—significantly frees up Rite Aid pharmacists' time to attend personally to customer needs.[15]

All of these examples share some key themes. The replacement of frontline service providers with machine-mediated services does not mark a transition to low-end, stripped-down, or second-class offerings because these machines consistently outperform the available human alternatives. In pharmacies, for example, the automated interfaces provide customers faster access to transactional services, while offering them greater convenience and privacy. At the same time, their access to pharmacists increases because pharmacists no longer allocate time to mechanical tasks. So, companies can not only realize labor savings and better efficiency, but also redeploy human labor to customer-pleasing activities that derive greatest leverage from their workers' uniquely human qualities, while capitalizing on the essential performance attributes of machines. As Mary Sammons, president and CEO of Rite Aid, said in an address to the National Association of Chain Drug Stores:

> Every time a pharmacist performs a routine task that a technician or assistant could handle better is lost productivity. Every minute a pharmacist spends cutting through red tape on the phone instead of talking to patients is a lost opportunity. We must strive to adopt standardized platforms and procedures so that we can communicate electronically, exploit robotics, accelerate distribution, improve workflow, and make better, more productive use of the personnel in our pharmacies. We must embrace what's new and available today, like e-prescribing, and envision what can be available tomorrow. . . . The goal, of course, is ultimately to free pharmacists to spend more time where it is most valuable . . . in direct interaction with patients.[16]

Productivity Parallels Between Manufacturing and Services

U.S. productivity gains were less than 3 percent a year from 1948 to 1973, falling to 1.5 percent from 1973 to 1995. From 1996 to 2000, productivity ran at 3 percent, rising to over 4 percent in the last two years and to over 5.5 percent in the last year.[17] Given the service sector's dominance in the U.S. economy, recent productivity increases inevitably point to the service sector. Today, front-office jobs are subject to forces similar to those that wiped out thousands of back-office jobs, especially in manufacturing and then data-processing, over the past half century: namely, the twin forces of labor displacement and automation. In industrial production, the elimination of jobs continues; since 2000, 2.8 million U.S. manufacturing jobs, or 16 percent of the total, have disappeared in the United States as work has shifted to offshore markets such as China and production automation has continued to make gains.[18] (Factory jobs now account for about 11 percent of all non-farm jobs in the United States, down from 27 percent in the 1960s.[19]) Similar patterns are emerging in the service sector.

When machines can occupy thousands of frontline jobs in service operations and lower-cost labor markets can perform many jobs remotely, the radical redesign of work is clearly not theoretical. Consider the fast-food industry. Individual franchisees of major brands have begun experimenting with methods to deliver better service at lower cost through their restaurants. As a major service channel, the drive-thru window has become the focus of reengineering efforts because ordering a meal there is such an awful experience. From an operating perspective, the problem goes beyond antiquated technology; drive-thru clerks are stationed amid chaotic and noisy food service facilities. So why not place these drive-thru operators in a remote, centralized call center so that they can manage order-taking for several restaurants at once? That's now happening market by market in local geographies. As franchisees aggregate such operations, we could soon be ordering dinner through our car window from someone in Bangalore or Leeds. Success in such front-office reengineering efforts will depend on driving both performance and cost to maximum customer advantage, not by diminishing one to realize the other. This goal—optimizing both performance and cost dynamics—requires strategic insight into how companies use displacement and automation as instruments of productivity gains and differentiation. As we will see, this view of the challenge resembles the first reengineering revolution that unfolded in corporate America some two decades ago.

FRONT-OFFICE REENGINEERING

Like the front-office revolution, the first reengineering movement was also spurred by the use of technology to automate office work. It, too, became a driver of productivity gains and, when successful, substantially improved the enterprise economics of companies that pursued it. It, too, created new possibilities for the design of work in large-scale corporations, promising gains measured in both efficiency and effectiveness. When Michael Hammer coined the term, he defined *reengineering* as "the radical redesign of business processes for dramatic improvement." The mantra of reengineering was summed up as: "Don't automate. Obliterate."[20] Hammer, Thomas Davenport, and other adherents argued that information technology in most large-scale enterprises had done little more than automate existing processes. Even if such imposition of technology could make a company's internal processes faster, better, and cheaper, it did not generate significant productivity gains if the processes were outmoded or inappropriate. The truth was, many corporations had established their operational practices decades earlier, with an eye to human, not machine, labor. That's why Hammer argued for fresh thinking about processes first and the application of technology second.

Reengineers typically focused on the back office, where mainframes and minicomputers had made substantial strides in automating data-processing functions, speeding the processes associated with record keeping, generating financial reports, tabulating market research, and running internal accounting systems. Despite the accelerating impact of machine labor, many corporate back offices were nonetheless hidebound places, full of multiple hand-offs, fragmented decision-making authority, and generally Byzantine process flows. Reengineering sought to eliminate these dysfunctional aspects of corporations, introducing new processes that brought people and machines together to maximize efficiency in performing a variety of back-office tasks.

As reengineering became a movable feast for consultants selling systems integration and organizational redesign, it also became a lightning rod for hostility among office workers who were reacting to the cold realities of corporate life. In the recession of the early 1990s, reengineering became inexorably associated with notions such as delayering, downsizing, and rightsizing corporate work forces. It took on a distinctively negative cast, as perpetuated to this day in Scott Adams's "Dilbert" cartoons. In fact, the central inspiration of reengineering was different. In its purest form, it was a theory for how to redesign internal corporate processes to take full advantage of what humans and machines could do best separately and together. It was not a program for the elimination of jobs. When reengineering worked, the potential for value creation was so great that it literally transformed many businesses, especially those with intensive data-processing activities such as credit card issuers and insurance providers, by enabling companies to incur significantly lower operating costs while delivering better results for customers.

Most of the reengineering success stories during this era were not associated with top-line growth. The focus was on cost compression and efficiency, but some benefits trickled down as a result of better processes to customers. In an era when fast food was slow, buying an airline ticket was complex, and credit card issuers did not answer their phones, being fast in fast food (Taco Bell), making complex airline reservations simple (Sabre), and answering phones unfailingly in two rings (MBNA) represented distinctive strategies. Capital investments in IT facilitated such competitive differentiation, even if business economists at the time argued whether corporate investments in IT had generated productivity gains.[21] However, corporate leaders who understood the implications turned reengineering insights into real sources of competitive advantage.

Reengineering in the 1980s was made viable by a convergence of technology evolution, which at the time involved different breeds of networks and devices from those today. The devices were mainframe computers operated largely by MIS professionals (the desktop PC, only a few years old, was just beginning to appear in some corporate offices) and private or so-called corporate by-pass networks and, in some cases, time-sharing systems. Like the front-office revolution, reengineering was based on new forms of processing power combined with new forms of network connectivity. But because these devices and networks were not diffused throughout the work place, they had little bearing on customer interactions or on the majority of employees in any corporation. These innovations belonged primarily to corporate headquarters and regional corporate data-processing centers; they seldom touched the majority of workers who interacted with customers on the front lines. Such distinctions notwithstanding, it's interesting to note that both versions of reengineering are concerned with a profound question: When machines enter the work force, how should companies and their managements adapt? In the 1980s, answers to that question focused on internally facing business processes involving interactions among managers and employees. Today, answers to that question focus on external business processes involving interactions among frontline work forces and the customers or markets they serve.

Elements of the Reengineered Front Office

The reengineered front office is composed of those interfaces a company deploys to manage interactions and relationships with customers. The three interface types are *people-dominant*, *machine-dominant*, and *hybrid*. We cite these as the three fundamental approaches to mediating customer interactions in work forces made up of people and machines:

1. The traditional approach to service delivery: people as frontline service providers

2. The automated approach to service delivery: machines as frontline service providers

3. The hybrid approach to service delivery: people and machines in combination as frontline service providers

To reengineer the front office, it is essential to understand the strengths and weaknesses of each of those interface types and to choose the most effective and efficient interfaces, or combination of interfaces, to mediate particular company-customer interactions. Of course, few companies deploy a single interface to the market. No bank could do business using ATMs alone without branches or call centers. No credit card issuer could provide cards without toll-free help lines and merchants who accept the card. Arguably, some Web-based businesses, such as Monster.com or Match.com, manage to do business largely with single hub or portal Web sites, but the sites themselves represent combinations of different types of interfaces, or pages, operating as a coherent system within the site. Companies managing interactions with customers generally accomplish the task through a variety of interfaces. Different customers will access a company's services in different ways, depending on their needs and usage occasions. That means companies must design not just interfaces but the systems within which those interfaces operate. The result of this process, which determines a company's interface capability with respect to its customers, is what we call an *interface system*. The interface system is ultimately the face the company puts forth to its customers and markets; an optimal interface system is a company's best face forward.

Let's examine each of these interface types briefly here; we will then explore their implementation in greater detail in chapters 4, 5, and 6. Figure 3-1 provides a visual overview of the types of examples we will discuss.

People-dominant interfaces

These interfaces represent the traditional approach to service delivery: retail store clerks, fast-food workers, retail bank tellers,

FIGURE 3-1

Interface Archetypes

	Dominant	**Hybrid**
People	**People-Dominant Interfaces** All in-person employees (Sales and service staff)	**People-Led Hybrid Interfaces** Employees with PDAs Employees with wireless headsets
Machines	**Machine-Dominant Interfaces** ATMs Web sites RFID tags and readers Vending machines Kiosks VRUs	**Machine-Led Hybrid Interfaces** Phone customer service Online chat help Broadcast television

Dominant: Interfaces consisting primarily of either people *or* machines
Hybrid: Interfaces consisting of a combination of people *and* machines
People versus machines indicates the interface type in direct contact with the customer

maids in hotels, waiters, flight attendants, doctors and nurses in hospitals, and attorneys in law firms. Each of these service positions is susceptible to change as machines enter the work place, but these are relationship management positions for traditional job categories and definitions. As such, they are primarily human interfaces with customers, or what we call people-dominant interfaces. These are the roles and activities of frontline service workers.

Machine-dominant interfaces

These interfaces represent a more recent development. Commercial use of machines in frontline service positions arguably began with vending machines and player pianos a century ago, but serious commercial intent arrived with the ATM, a device that enabled retail banks, starting in the late 1970s, to substitute capital equipment for labor on a widespread basis across branch banking systems. Today, machine-dominant interfaces are increasingly ubiquitous, including in the online world of e-commerce sites. In the offline world, they are equally pervasive in kiosks, vending machines, talking elevators, and voice-activated automobiles and phones. While every one of these devices is supported by human beings, who maintain the hardware and program the systems and services, machines largely perform interactions with customers; that's why we call them machine-dominant interfaces.

Hybrid interfaces

Hybrid interfaces are more complex. These are interfaces wherein the presence of both people and machines at the point of interaction is evident and palpable to customers. Some hybrids may place people in the foreground and machines in the background, like a travel agent consulting a reservations system while booking a vacation. Others may place machines in the foreground and people in the background, like an airline's electronic ticketing machine with counter personnel available to check bags. In this sense, there are *people-dominant hybrids* and *machine-dominant hybrids*. Both fall into a category that involves an amalgam of human and machine attributes.

The three building blocks of the reengineered front office are archetypes. Every interface a company deploys will likely involve some level of human and some level of machine intervention. The art of determining the appropriate interface to deploy at any given point of interaction between a company and its customers is something we'll discuss in detail later. For now, it suffices to say that the choice of interface type involves critical trade-offs between segment-specific customer needs and expectations on one hand and a company's capabilities and resources on the other. And while companies must make difficult choices one interface at a time, they must also take into account interface system effects. In most businesses, interfaces seldom operate in isolation from one another. That means that a critical challenge in front-office reengineering is not only optimizing individual interfaces on a stand-alone basis but also optimizing them, in concert, as a system. System considerations may, in turn, alter the choices a company makes about individual interfaces, since these two variables—individualinterfaces and interface systems—must exist in dynamic equilibrium over time.

Systems Thinking

As technology has facilitated new types of interfaces between companies and customers, strategic thinking about interfaces in systems has evolved dramatically. Just as we expect most companies to have multiple channels of distribution, we assume that they will deploy multiple interfaces to customers in their markets. Managers, however, have not always harbored expectations of multiple channels and interfaces. For example, in the late 1980s, a research group at Harvard Business School interviewed John Reed, the CEO of Citibank, long before the merger with Travelers. At the time, PC home banking was an emerging innovation, and Citibank, along with Chemical Bank in New York, was leading the charge. Citi's Home Banking service and Chemical's Pronto System were high-profile attempts to persuade customers, who had just begun using PCs in the mass market, to bank online—when getting online meant using proprietary dial-up lines to connect directly with the bank's mainframes. Though Chemical ultimately lost some $200 million on Pronto, and Citi persisted for years thereafter, Reed believed that automation of retail banking services nevertheless represented the industry's future.[22] When asked what kinds of changes the new electronic channels for banking services portended for the branch banking business, Reed quoted some statistics: The fully loaded cost of serving an average consumer retail account in a year in Citi's New York branches was roughly $500. He estimated that the annual cost of maintaining a customer account through Citi's call centers at the time was $100; through its ATMs, it was $50; through dial-up, it was $25. Draw-

ing the dramatically sloping cost curve in the air, Reed pointed to the downward trajectory and asked, with a rhetorical flourish, "Where do *you* think the banking industry is heading?"[23]

In one sense, Reed was right. Especially now, with the advent of the Web, it's undeniable that automated interfaces such as the ATM and the PC represent the lowest cost for retail banking services. It's also undeniable that customers would ultimately embrace these new channels to access banking services. In another sense, Reed was sorely mistaken in his assumptions, which is clear only in light of subsequent history. The flaw in Reed's thinking was he assumed that each lower-cost service interface would provide a pure substitute for the one that had come before: call centers would replace bank lobbies, ATMs would replace call centers, and PC-based banking would replace ATMs. From a cost perspective, substitution of this kind represented a grand and compelling vision. To make it a reality, in the 1980s Citi began charging customers for interactions with tellers while offering the comparable services free at ATMs. Of course, consumers objected. Customers seeking different interfaces with banks would do so based on their individual preferences and needs, which would vary from one customer segment or usage occasion to the next. For example, while getting cash may bring a customer to an ATM, an error in a monthly statement generates a call—usually after business hours—to a call center. What history shows in retail banking, like other service sectors, is that no new interface type has rendered a preexisting interface type obsolete, any more than one new medium (e.g., television) has caused any other (e.g., radio) to become extinct. After all, outbound telemarketing and e-mail have not eliminated direct mail; and online retailing has not eliminated shopping malls; because each service interface addresses unique customer needs and occasions. That's why, in retail banking today, machines have come to mediate 59 percent of all customer interactions, but this figure represents a combination of the ATM, phone, and online transactions.[24]

The conversation with John Reed illustrates a profound point for anyone who would tackle front-office reengineering. As we learned in the 1980s, reengineering is a double-edged sword. Executed well, it can reduce a company's costs and improve performance. Executed poorly, it can raise costs while degrading performance. The Citi formula was one that would, in effect, trade service quality for anticipated cost savings. The theory was compelling, except that the bank's customers did not stop using the branch lobbies, the call centers, or the ATMs. The net result of adding online banking was higher overall cost and complexity in operating the business, with increased need for service from a customer population confused by the additional channels and options. In this sense, merely proliferating interfaces to customers is not a sound strategy for cost compression or service enhancement. Only when interfaces are deployed selectively and thoughtfully, according to their optimal archetypes, and then managed strategically as systems do they result in a virtuous cycle of reduced costs and improved customer outcomes. This point has sobering implications. Why? Because a large number of companies have deployed countless interfaces to connect with consumers but optimized none of them and, worse, failed to ensure that each interface is integrated systemically with the others. Citibank, a pioneer in experimentation with retail automation, took years to work this out. But no company today has such luxury of time—and none must repeat these errors. The critical challenge is simple to state, complex to tackle. Every new interface a company deploys raises the costs of doing business by incurring costs incrementally, while it raises the overall costs of doing business by increasing costs associated with interface systems complexity. As a result, the selection and deployment of each interface must trade performance for cost, while the overall interface system must optimize ultimate trade-offs between effectiveness and efficiency. Front-office reengineering matters because it represents a process whereby companies can increase revenue growth while driving down costs of providing service, but it demands artful management to get there.

Old Way Versus New Way: Envisioning
the Reengineered Front Office

Let's digress briefly to illustrate these ideas with some practical realities, by examining an age-old industry in which it's relatively easy to compare a traditional front office and a reengineered front office. We'll spend some time with gaming. If you're in the principality of Monaco, you might be tempted to walk into the casino off the town square in Monte Carlo. There, you will see an environment for games of risk that encapsulates what gaming used to be—a palatial space with an expansive array of roulette games and baccarat tables. Not only is the Monte Carlo casino lavishly appointed in Baroque grandeur, but service staff and patrons dress in dinner jackets or formal wear. At each table, a croupier leads the action amid solicitous assistants, supervisors, and waiters who proffer drinks to the seated crowd. This approach to interface capability may involve very high operating costs, but catering to the upper crust of high rollers and high-net worth individuals yields equally munificent rewards. Not surprisingly, the scene has not changed substantially in a hundred years.

But no casino in the United States resembles Monte Carlo. The people-dominant model of running a casino's gaming floor is gone, replaced by magnificent arrays of interface systems composed of all three interface archetypes. The gaming industry's equivalent of the ATM is the slot machine. Automated teller machines and slot machines both began as electromechanical devices decades ago. (The first cash dispenser, introduced by Barclay's in the United Kingdom in 1967, utilized punch cards and rubber bands to distribute cash.[25]) Both have evolved into sophisticated, focused-use machines that harness digital technology, high-resolution color screens, and ergonomic designs to enable interactions that please their habitual users. Like ATMs, slots don't require a particular retail environment to attract users; their owners can place them wherever consumers' needs or desires might legally be fulfilled. (Consumer research in retail banking has shown that bank branches are not among the six most popular locations for ATMs.[26]) Hence, in Las Vegas or Reno, you see slots in hotel lobbies, airport concourses, grocery stores, and restaurants and bars, often alongside ATMs (a potent pairing). Such machines are everywhere you want them to be.

For the gaming industry, the lowest-cost channel for service delivery is the slot machine, the industry's only fully automated interface besides online betting. One variant of a slot machine game, video poker, qualifies as the single most addictive form of wagering in the history of the gaming business.[27] Since gaming operators can set the odds on each slot machine, their attractiveness from a business perspective is obvious: Not only are slots low-cost interfaces, but they also reliably deliver whatever profit margin their owners choose.

Consider that dynamic from the perspective of one casino operator, Harrah's Entertainment, which runs twenty-six casinos in twenty markets constituting the most widely distributed gaming facilities in the United States. At each location, Harrah's operates hotels and restaurants, and all manner of theatrical and musical entertainment at its major properties; and it employs more than forty thousand people, most of whom work in service jobs in over 1.5 million square feet of retail space. Despite the scale of its human organization, Harrah's generates over 80 percent of its profits systemwide from its slot machines.[28] These forty-two thousand machines process roughly $50 billion in wagers a year.[29]

Why would Harrah's want to operate such complex retail environments when its automated front-office labor is so efficient and effective? Why would *any* gaming company bother with complex casino operations when slot machines can attract, retain, and frequently mesmerize customers? Because optimal customer interactions rarely derive from interface systems composed of a single interface archetype or one devoid of people. The entire labor-intensive context of Harrah's casinos, and not simply the slot machines themselves, attracts $50 billion in wagers every year. Using slots in this manner differs from retail banks' experiences with ATMs. The widespread deployment of ATMs starting in the

1980s created breakthrough convenience for customers but barely differentiated or generated excitement around individual banks or brands. If anything, ATMs homogenized banking services, especially once ATM networks became interoperable across banks, while depersonalizing customers' retail banking relationships. The success of ATMs led to their dominance in consumer banking relationships, but that dominance became a central problem. Single-interface systems do not serve customers with varying needs and usage occasions well (unless the interface, like a well-developed Web site, is itself an interface system), and basing a distinctive or branded relationship on them is often difficult.

Harrah's serves customers on its casino floors in two ways: through people-dominant interfaces at its table games (Monte Carlo-style) and through machine-dominant interfaces at its slot machines (ATM-style). For gaming companies, deploying multiple interface types drives revenues higher while lowering costs. Here's how: The action on the casino floor—around table games such as craps, blackjack, and baccarat—makes the casinos exciting. To generate traffic, casinos need action; to generate profits, they need slots. In a sense, all frontline workers—including the dealers, waiters, and bartenders—work the floor to sustain the electricity of continuous play. In Monte Carlo, people alone can provide the interfaces because the target market is small and margin opportunities are large. But in the consumer mass-market of gaming that Harrah's targets, optimizing the interface system requires people and machines, strategically operating together to drive play most effectively while generating revenues most efficiently. Since gaming companies can determine their own odds across most of their games, not just slots, they effectively invest in labor-intensive services to attract the levels of traffic needed on their casino floors, which drives revenues and then generates profits at predetermined rates.

So that's why Harrah's manages front-office reengineering as a continuous process. Its hybrid interfaces involve liberal experimentation with Wi-Fi devices. By giving attendants at Harrah's properties handheld devices to register guests as they arrive, the company can check elite loyalty-card holders into its hotels at curbside, helping them avoid long lines at reception desks, resulting in players with higher levels of satisfaction and more time to wager. Such player identification is facilitated by the Total Rewards program, which has data on the habits across Harrah's properties of some 25 million customers. The program relies on a combination of database marketing, yield management, and decision science. Drawing on loyalty-program data, the handhelds can cue employees to offer arriving guests complimentary meals, hotel rooms, or tickets to shows, depending on their status as Harrah's gamers. The company has also experimented with roving cashiers, who use handheld computers to cash players out on the casino floor rather than requiring them to stand in lines at cashiers' windows, and with portable printers alongside table games to generate federal tax forms.[30] The intent of all these experiments is to maximize play among the company's most active gamers while incurring the lowest cost of service.

Monte Carlo illustrates gaming before front-office reengineering and Harrah's illustrates gaming after front-office reengineering. The two casinos have created very different worlds, with radically different implications for the companies that operate them. Monte Carlo creates excitement with personal intimacy and recognition for high rollers, but there are limits to the scalability of its model. It works best as a kind of private bank for a small number of high-net worth individuals. Harrah's furnishes excitement with database-driven intimacy and recognition, but there are potentially no limits to its scalability. Its interface system is engineered as a kind of Disney World of gaming for a vast consumer mass market. Only an interface system built of hybrids could combine machine intelligence and personal service in this way. In Harrah's casinos, every slot machine and table game has a swipe-card reader. That means data on any individual's play is captured in real time, with networked systems making inferences from the data available to floor personnel using display devices as they are interacting with each player. This behind-the-scenes information capture creates the potential for manufactured intimacy in a mass-market environment. That intimacy can be used to adjust the pace or odds of play across multiple games, table or slots, and distributed across multiple properties as players move across the Harrah's system. The alchemy is the result of all three interface archetypes operating in an integrated interface system that capitalizes on the strengths of people, machines, and people combined with machines. Theoretically, in a small venue like Monte Carlo's, frontline service providers of high quality can remember the names and preferences of frequent visitors, adjust the drama and style of games to suit individuals' needs, determine the odds for the house based on well-honed instinct, and reward loyal patrons over time. But no one can do that through people alone in a mass-market environment. At Harrah's, the reengineered front office has resulted in stand-out growth and profitability. Harrah's furnishes its players with Main Street intimacy and its shareholders with Wall Street wealth.

No wonder retail banking companies are now playing the same game. As ATMs and Web sites have reduced the need for brick-and-mortar bank branches and, with them, retail opportunities for differentiation, some banks have converted their branches into more diversified environments to capture excitement akin to that of casino floors. Wells Fargo, for example, retrofitted redundant branches as customer cafés, where its account holders can drink coffee, relax, and interact with bank personnel or access financial information online.[31] E*Trade extended its operations from its Web site to its own retail stores that blend human and machine interfaces.[32] ING Direct—which offers a concise line of savings-oriented products with higher-than-average yields online and by phone—has created a physical retail presence consisting of ING cafés in select cities around the world.[33] These shops, sporting the same bright-orange color of the company's Web site and collateral materials, may not perform many banking functions, but they use various interfaces to differentiate an otherwise generic collection of commodity services around which ING has built a powerful brand. The cafés signal no-nonsense simplicity (in a complex category), excitement (in a low-engagement category), respect for the customer (in an impersonal category), and a results orientation (in a category famous for putting banks' priorities ahead of customers'); they also deliver a modicum of sex appeal, and, in the gaming sense of the word, "action." Interestingly, ING Direct's initial business plan did not include the cafés, which are few in number and located only in major urban centers. When ING Direct (which launched Canada as its first country market) found that customers were coming to its call center outside Toronto to visit the people who held their money and confirm that the bank was real, the company developed and implemented the café idea, deploying a new interface in Canada and globally to meet that customer need.[34]

Interfaces and Service Management

As interfaces mediate interactions with customers, one might wonder: What functions, exactly, do interfaces perform? Interfaces deliver services. Service is how most companies describe the ways in which they interact with and relate to customers. When an ATM delivers cash, the machine is performing a role normally associated with a frontline service provider. As a result, it's important to underline that the functions or activities of the front office are almost universally service related. This compels us to conclude that any interface deployment is subject to the lessons and wisdom related to managing in the service sector. Whether the company in question is an industrial or a service organization, and whether the interaction is delivered by a human frontline service provider or by a machine, the challenge is one of service management. Let's consider the literature.

Serious research in services originated with the conviction among business scholars that the challenges of managing service businesses were different from those of industrial businesses. Because service interactions traditionally involved face-to-face encounters between frontline service workers and customers, orchestrating successful service operations was predicated on human attitudes and behaviors in ways that had no parallel in industrial operations. These unique managerial challenges involved an understanding of how people, processes, and systems interacted to field vast work forces of employees who directly touched a company's customers. Moreover, unlike manufacturing businesses, service companies cannot inventory their outputs in advance or check them for quality before they're delivered to customers. Generally speaking, services are created through, or as a result of, interactions between service providers and customers. (Customers cocreate some services, such as education and training, and customers trigger the delivery of other services, such as the preparation of a restaurant meal.) Moreover, effective service providers change what they do based on customer interactions, so they need degrees of freedom and latitude to get their jobs done; this makes command-and-control approaches to service management often unworkable, despite attempts to codify approaches to service delivery based on detailed manuals and specifications.[35]

One model for a new approach to service management came from our colleagues at Harvard Business School. They called it the *service profit chain* (and, more recently, the *value profit chain*).[36] The service profit chain codified a set of causal relationships in service businesses that linked employee satisfaction and loyalty, customer satisfaction and loyalty, and positive financial outcomes. In so doing, the model established criteria for identifying successful service businesses as those that optimized outcomes for three constituencies: employees, customers, and shareholders. More satisfying and meaningful jobs that result in *employees* becoming more loyal to, and productive within, a firm constituted a successful outcome. Better service that yielded more satisfied and loyal *customers* constituted a successful outcome. Greater revenues and profits above sector averages for shareholders also qualified as success. Each of those successes was, by no means, as extraordinary as a company's achieving all three objectives at once.

Only a few companies manage themselves to optimize results for employees, customers, and stakeholders simultaneously. While this outcome may seem an obvious goal, success in practice is rare.

Research on managers' day-to-day priorities shows that they routinely trade off one constituency against another to meet short-term goals. For example, if management realizes that it must generate more profits in any given quarter, it could give less to customers—less attention, fewer amenities, and longer wait times—or less to employees—lower pay, fewer bonuses, and no promotions. We all know the consequences of such thinking; we experience them daily as consumers. You return to a favorite hotel only to find that the quality of amenities has diminished. Or you return to a favorite restaurant for its attentive service, to find that each waiter is handling (and rudely so) too many tables due to staff cuts. According to conventional wisdom about customers' perceptions of service quality, performance must meet or exceed expectations for customer satisfaction and loyalty to increase over time. Consequently, reducing service levels to boost financial returns may benefit short-term financial outcomes but endangers future revenues.

A diminished quality of interaction between a company and its customers erodes the competitiveness of a business with customers, and it diminishes the job satisfaction of employees. Generally, the attitudes of employees mirror those of customers; through the quality of their interactions, satisfied and productive employees produce satisfied and loyal customers. But the operating decisions of many managers prove the converse to be true: the higher the manager in an organization, the greater the commitment to service quality; the lower the manager (and, ironically, closer to the customer), the more financial goals take precedence. So consumers experience a discontinuity between what organizations say and what they do. Recall those slogans from the airline industry, such as US Airways' "USAir Begins with You," Delta's "We Love to Fly, and It Shows," and United's "Fly the Friendly Skies"? These tag lines all delivered a consistent message: Even though we have the same seats and the same jets, we are different—because of our service! When you show up for your flight, however, the rhetoric seldom comes to life. Most of us have learned to expect that frontline service workers who deliver on a corporation's advertised promises are the exception, not the rule. Yet the truth is that airlines, like most businesses, establish competitive advantage first and foremost through the quality of their interactions with customers.

Why would smart companies operate against their own competitive interests? Worse, by raising expectations through marketing programs, they are disappointing customers by a wider margin. In the trade-offs between short-term cost reduction and long-term customer satisfaction and loyalty in the long term, the long term usually loses. Businesses seeking advantage must find ways to stay the course—to invest strategically in the quality of their customer interactions. The problem, however, is multifaceted: Customers are more demanding than ever because they have unprecedented levels of choice and transparency in markets. Leading companies, by improving customer relationship management, have raised the bar. And, most important, merely incremental gains in customer satisfaction in interaction with a company usually fail to deliver enhanced customer loyalty. For businesses in competitive markets, customer attitudes (satisfaction) must shift decisively in a positive direction to result in meaningful changes in customer behavior (loyalty).[37] For years, companies have used 5-point scale surveys to track service. Many managers at such companies have assumed that merely good, not great, ratings are good enough, defining thresholds of success at, say, a 3.7 on the 5-point scale.[38]

In competitive markets, however, only extremely high levels of satisfaction command the attention of consumers and induce them to change their loyalty behaviors. The reality is that satisfaction must rise to extraordinary levels in most businesses before prospective customers will curtail their propensity to shop across available alternatives. The hurdle rate for behavior change of this kind is satisfaction levels of 4.5 and above on a 5-point scale. As companies reach the upper end of the scale, loyalty becomes manifest in a variety of related virtuous effects, including positive word of mouth, higher referral rates, diminished price sensitivity, greater volumes of business from each account, and lower costs of new account acquisition.[39] These effects add up to the financial returns in the long term that result from getting customer interactions and relationships right. (Under monopoly conditions, these observations obviously do not apply; some market-dominant businesses—e.g., Microsoft, with the Windows operating system—had, until recently, the luxury of taking customer satisfaction less seriously.)

Such virtuous effects resulting from material change in attitude and behaviors help companies in competitive markets compete for depth of customer relationship (share of wallet) and breadth of customer reach (share of market). From a strategic standpoint, managers find these ideas compelling, but the monthly or quarterly plan often trumps the longer-term objectives such as increasing customer lifetime value or market share. In today's competitive environment, companies in most industries cannot merely exercise cost leadership (operating efficiency) or product differentiation (effectiveness). They must do both. Now that interactions are the new competitive frontier, companies must serve customers in ways that meet short-term cost objectives *and* long-term objectives for quality of interaction.[40]

Front-office reengineering does exactly that. It uses new forms of technology-enabled substitution (automation) and network-enabled displacement (off-shoring or outsourcing) to compress the costs of deploying service interfaces while enhancing their quality and performance. The literature on services indicates that managers should not underestimate the effects of such approaches. For example, one research study completed several years ago measured the role of service quality in the decision-making processes of customers who had recently switched vendors. The variables the researchers examined were service quality, price, product features and functions, and availability. The results showed that service quality had five times more weight in influencing purchase decisions than any other attributes tested.[41] This indicated that the critical component affecting customer purchase behaviors is the quality of interaction with a company, over and above price or performance.

MACHINES AS SERVICE PROVIDERS

Because critical building blocks of the reengineered front office are machine-dominant interfaces, managers must subject technology-mediated interactions to similar kinds of scrutiny. How well does a machine perform as a service provider or relationship manager? Does the machine satisfy its users? Does it satisfy them enough to change loyalty behavior? Obviously, a Web site that interacts with users in satisfying ways will likely produce loyal customers. (How else do you explain the extraordinary staying power of Amazon's or eBay's millions of active customers?) So we can reasonably believe that elements of any interface system—whether people-dominant, machine-dominant, or hybrid—must operate with the same goals that might govern a well-managed traditional service delivery system. Be it human or machine, every service interface must function so that it delivers higher levels of customer-perceived value relative to competition, resulting in customer satisfaction and loyalty rising sufficiently to drive superior financial returns.

Consider a metaphor from the software world. When programmers develop software packages or Web sites, they sooner or later face the challenge of designing the screens that their software will present to users. They must ask themselves: How will this software interact with the user? How will critical information be organized and presented? How will the user activate specific features and functions? Getting even more specific, they must address questions of screen design: What will the home page or log-in screens look like? What icons will users click on to activate certain functions or services on-screen? How simple or complex should each page be? What kind of aesthetic appeal should the screen designs have? This aspect of software development is often referred to as the *presentation layer*.[42] The *interface system* is, in effect, the presentation layer of a company; it presents the content, functionality, and services of a company to its users, its customers. Like software and systems, from a customer or user perspective

companies are worlds unto themselves, often opaque, difficult to
navigate, and subject to their own internal rules and idiosyncrasies
governing what they do and how they behave. To render a company accessible to customers, it must have a presentation layer; the
alternative is that customers must reach down into the code to
make things work (which works to some extent in cases such as
self-service restaurants and warehouse clubs, where the code becomes the presentation layer).

The presentation layer is an apt metaphor for understanding
the interface system of a business, especially in light of recent findings in the fields of psychology, neurosciences, and aesthetics. Recent research indicates that positive interactions with both service
providers and physical products can win goodwill through enhanced usability, design or aesthetics, and reward or recognition.
For example, Alice Isen has shown that interactions with human
subjects following an intervention designed to create a positive
emotional state (an unexpected gift of a dollar or a chocolate bar
given to a research subject) generate a disposition more accepting
of flaws and infelicities. One might argue that such research findings explain the commercial value of Wal-Mart's greeters or Amazon's logo that incorporates a smile. Isen shows that such positive
interventions produce traceable changes in neurotransmitters resulting in a literally altered state of mind. From a different angle,
Don Norman in *Emotional Design* describes how "attractive things
work better," in that users will more likely overlook minor problems in design and tolerate "minor difficulties and irrelevancies."
Superior interactions through customer-satisfying approaches to
service interactions affect customers similarly, regardless of how
those services are mediated, resulting in faster decision making,
greater likelihood of attachment, and elevated levels of empathy
and forgiveness.[43] Finally, and more recently, Virginia Postrel has
argued that aesthetic dimensions of products and services can
engender similar positive effects, suggesting that companies can
create value through "sensory" and "expressive" attributes of offerings, not just through "intellectual" or "physical" attributes.[44]

As companies compete increasingly on the quality of their interactions with customers, the psychological impact of their interfaces has ever greater strategic import. Just as many a Web business lives or dies based on the appeal, ease-of-use, and functionality of its presentation layer to customers, companies must view their interface systems as critical sources of strategic positioning and competitive advantage.

CONCLUSION

In this chapter, we have argued that reengineering the front office follows a logic similar to that which drove the first reengineering revolution some two decades ago. The differences between these two waves of reengineering, however, are important. In the 1980s, reengineering focused on the internal world of corporations, generating efficiencies in the corporate back offices. The front-office revolution is about how companies interact with the external world of customers and markets. In examining the contrast, we have touched on a number of themes:

- *We are entering an era where front-office applications of technology are changing the strategic possibilities for companies in their largest job category—services.* Just as automation came to agriculture and manufacturing, front-office automation is finally coming to services. Whether machines substitute for frontline service positions or networks connect those roles to jobs in remote geographies, technology in the front office is creating the potential to radically reconfigure work in the largest sector of the economy.[45]

- *Three types of interface constitute the building blocks of the reengineered front office.* Few businesses can rely on a single interface to manage interactions with customers and markets. Most companies have already deployed a variety of interfaces to customers for reasons of accessibility, convenience, and competitive parity. Many companies are already making use of people, machines, and people in combination with machines. Reengineering the front office uses innovative combinations of people and machines to radically redesign customer interaction and relationship management functions to build cost *and* revenue advantages.

- *Front-office reengineering demands that we rethink and reframe our notions of what it means for a company to go to market.* New technology enables companies to deploy interface systems that express the essence of their differentiation and brands. But this promise is only realized where companies and their leaders think broadly and creatively about how people and machines, as equal partners in frontline labor, manage interactions and relationships with customers.

- *Whether people or machines manage interactions with customers, lessons from the service sector apply.* What the service management literature highlights regarding how companies compete on services is broadly applicable to interfaces and interface systems. Machines in service roles are analogous to human service providers. Success for machines and people must be assessed in similar ways, as measured by customer perceptions of value and resulting levels of customer satisfaction and loyalty. The notion that interfaces can have profound impacts on how customers perceive value in companies is supported by a growing body of scientific literature. This research indicates that customer perceptions are strongly influenced by the emotional context in which interactions take place. In the original reengineering literature, proponents argued that companies must aspire to become "easy to do business with." Today, companies must deploy interface systems engineered for efficiency and effectiveness in delivering on that goal. We examine the first of the three building blocks of such systems in chapter 4.

4

WHAT PEOPLE DO BEST

IT'S ALMOST QUAINT in today's world to consider businesses where people work largely with people to manage interactions and relationships with customers. Such businesses are the rare but persistent carry-overs from the traditional world of services. Yet, in some contexts and markets, they can still prove successful. For example, along the freeways in Los Angeles you can find a thriving fast-food chain called In-N-Out Burger, a brand that built its success on the drive-thru window. Each In-N-Out Burger looks like a restaurant from the 1950s. When you're in line for service, you may see employees in In-N-Out uniforms running to and fro carrying crates of shredded lettuce and burger buns, but there are no touch screens or PDAs or swipe-card readers. The chain easily qualifies as the source for one of the best fast-food hamburgers in North America. In-N-Out recently tied for first place in customer satisfaction among fast-food brands in a survey of over sixty thousand consumers throughout the United States.[1] But other than some cooking equipment and an electronic cash register, the chain uses little, if any, front-office automation. In contrast, many an underground parking garage has machines like those found just off Chicago's Lakeshore Drive. When you return to the garage to get your car, there are no people behind the cashiers' windows (though the windows are still there), only signs pointing to machines across the lobby. The machines process the parking payments, using a combination of visual commands and on-screen and voice prompts. We gave our ticket to one machine; it spoke back with a male voice produced by an Asian synthesizer chip with a faint South Side Chicago accent. It asked for a credit card, which we proffered; it processed the transaction; and returned the card and the ticket, directing us by voice to our car. On exiting the garage, a machine took the validated ticket, raised the gate, and thanked us for our business.

What a contrast to experience both In-N-Out's all-human and Lakeshore Drive's all-machine service interfaces! Yet both approaches deliver effective and efficient interactions to customers. Quaint *and* futuristic, people-dominant and machine-dominant interfaces coexist today—and each interface represents an important instrument of strategy and management, either as a stand-alone interface or in combination with an interface system, in establishing a company's competitive advantage. In this chapter, we examine the first of our three interface archetypes—the people-dominant interface. There's one caveat: Interactions and relationships with customers vary by industry sector in complexity and goals. Most companies aim to deploy interface systems that mediate interactions and relationships in unique and differentiated ways. Some interfaces within such systems may be generic on a stand-alone basis (ATMs and slot machines); others may be specialized or customized (Nordstrom salespeople and Amazon's Web site). As companies strive to design and deploy interfaces that push the frontiers of differentiation, individual interfaces (not to mention systems) can become complex. In this respect, we call our three interface types archetypes for a reason: they are the theoretical building blocks of which interface systems are composed. Which interfaces companies actually deploy may prove more complex than these three archetypes suggest, involving people supported by machines or machines supported by people, which, in turn, are supported by more machines and people. We'll explore such multilayered service interfaces later in the book. Most companies go to market, however, with a dominant interface type—what we call an *anchor interface*—to represent their brand, even in the context of complex interface systems. This selection, in itself, represents a critical strategic choice. For example, Nordstrom has chosen a people-dominant interface in its department stores in order to differentiate its offerings based on person-to-person interactions; Amazon has chosen a machine-dominant interface online to differentiate its offerings based on database-driven personalization and intelligence.

People-dominant interfaces invite us to consider examples from the landscape of traditional service businesses. We will revisit several exemplary service companies, but we'll do so from a nontraditional angle. Because we argue that people and machines perform comparable functions as interaction and relationship managers, we will apply the same lens to people as we would to machines in considering their readiness for work on the front lines. Our discussion of technology evolution in chapter 2 highlighted four attributes that qualify machines to begin to assume relationship management or front-office roles. We can restate those four attributes in human terms: physical presence and presentation (proliferation), cognition (intelligence and interactivity), emotion or attitude (affect), and social networks and interpersonal skills (ubiquitous connectivity). These are the traits that render humans well equipped to interact effectively with other humans. And these are the dimensions along which we must consider the division of labor in frontline work between humans and machines. Depending on the setting, the customer, and the service attributes most highly prized as a result, one profile of relative attribute strengths may prove more persuasive than another, favoring people, machines, or a combination of both. In this chapter, we begin our consideration of the three archetypes by examining the traditional service interface—humans as frontline service workers—to illustrate how interface systems operating largely without benefit of technology can, even today, deliver customer-satisfying interactions in optimal ways (see figure 4-1).

Let's start by illustrating the four attributes in action in people-intensive service settings. It's not unusual for businesses that

FIGURE 4-1

Relationship Drivers

PARALLELS IN RELATIONSHIP MANAGEMENT

Machine drivers of relationship management (Chapter 1)		Human drivers of relationship management (Chapter 4)
Proliferation	➡	Physical presentation
Interactivity and intelligence	➡	Cognitive abilities
Affect	➡	Emotional abilities
Connectivity	➡	Synaptic connections (in social and community contexts)

compete on the quality of interactions with customers to apply criteria based on our four attributes in recruiting talent into front-line positions. How do companies do this? They often seek or specify certain *physical* attributes of the people required (how workers should dress, behave, speak). For example, two financial services giants—Washington Mutual and Bank One—recently instituted dress codes for white-collar workers in their service operations. Their goals are the same—to field frontline people with a consistent physical self-presentation to customers.[2] Managers also specify *cognitive* attributes they believe are important or essential to success in frontline positions (creativity, intelligence, and problem-solving ability, or specialized knowledge). For example, investment banks and consulting firms are increasingly making use of case-study or problem-solving methods to screen and recruit new hires, using these techniques to test candidates against the intellectual challenges they will encounter in their prospective jobs.[3] Managers frequently consider the *emotional* or *attitudinal* traits that will prove most effective in a frontline role (empathy in health care, style in retail, stoicism in loan evaluations, warmth in hospitality). For example, chain retailers in fashion apparel, such as Abercrombie & Fitch, Gap, and Benetton, instruct their staff members to keep an eye out for potential hires among customers coming into their stores, aiming to recruit those who exemplify the attitude, good looks, and panache of their brands.[4] Finally, managers often prioritize recruits with *social networks* (the personal community of contacts and friends they may tap) and *interpersonal skills* (how they relate to peers, teams, or departments) that they may bring to the company. For example, one of the long-standing hiring criteria at Yahoo! has been "spheres of influence," people who have the ability to attract others like them to whatever they're doing—and thus bring more of the right talent into the business.[5]

The parallelism here between human resource requirements on one hand and technology attributes on the other is no coincidence. It's the result of our applying a common lens to the question of what makes people and machines fit to interact with customers in frontline positions. There is no news here when it comes to people. Having bodies and souls, hearts and minds, and social ties is what makes us human. The headlines pertain to machines beginning to manifest these attributes—or, at least, creating the perception that they possess such qualities. As we will see in chapter 5, this is why people are forging powerful relationships with machines—intelligent or appealing products of all kinds, not just service interfaces—in ways they never have before. In this chapter, however, the comparability between people and machines is essential to grasp. Without an understanding of the similarities between what people and machines bring to frontline service roles, we cannot begin to address from a managerial perspective the appropriate division of labor between the two in rigorous fashion. As a result, we may evaluate here practices among traditional people-dominant interface systems according to these criteria and explain why some are successful—and others are not.

PEOPLE-DOMINANT INTERFACES

In the early twentieth century, when managerial capitalism was a new idea, people-dominant approaches to services were the model for nearly all service businesses, ranging from the clerical (food service, hospitality, transportation) to the professional (banking, insurance, brokerage). Both clerical and professional service roles were typically extraordinarily labor intensive, often in less than stimulating ways. For example, it's remarkable to think back to corporate life in the 1940s and 1950s, when white-collar workers earned their keep dealing with repetitive, mechanical information-manipulation tasks to generate corporate financial statements, invoices, and reports. If they relied on technology at all, they used adding machines or fancier electromechanical devices called "bookkeeping machines." With the advent of mainframe computing and punch card data entry, many of those jobs changed or disappeared, eliminating what were massive human data-processing operations and replacing them with machines designed to perform similar tasks. Similarly, the advent of spreadsheet software for the PC in 1978, with innovations such as VisiCalc leading to blockbuster products such as Lotus 1-2-3 and Microsoft Excel, exerted a profound influence on how young financial and operating professionals in corporations, consulting firms, and banks spent their time. For example, in the mid-1970s, at management consulting firms such as McKinsey & Co. and Arthur D. Little, younger professionals with M.B.A.'s spent months and hundreds of thousands of clients' dollars in teams developing spreadsheet models, which were comprehensive financial analyses testing a variety of future options in a client's business. Today, in such firms, models of this kind are generated by a couple of junior analysts with limited business training in a matter of days, thanks to spreadsheet software running on laptop PCs.

Service on a Local Scale

It's easy to forget how significant basic information-processing tools have proven in automating information tasks for professionals. For white-collar workers, such tools have changed the very nature of work—and have often freed humans to think more broadly and conceptually in the process. While such automation has had its impact on the service professions, more mainstream service businesses have evolved much less dramatically, at least until recently. Consider a typical local mom-and-pop retail store in a city, suburb, or small town. Not much has changed in the ways such retailers operate in the past hundred years. While you may find a computerized cash register out front or a PC in the back office, mom-and-pop stores are people-dominant operations, with humans supporting humans. People constitute the interface system—each employee provides a physical interface mediating customers' interactions with the store. Of course, clerks also directly or indirectly manage other interfaces in the retail environment; they array merchandise, apply prices, and determine the store's look and feel. If owners hire teenage kids to work in the stores, there are no information systems to control their work; as a result, they require direct managerial guidance and support. Any attributes of cognition and emotion in the retail environment are people based, and any shared knowledge is passed along among members of the interface system person to person via word of mouth or, as some like to say, a sneaker-net.

If you ask for products or out of the ordinary information in such stores, you will see the sneaker-net in action. Inevitably, junior personnel seek answers from the owner about anything more advanced than a question about store hours. The last time you called such a store in your neighborhood with an unusual question (like many small-town children, we spent idle hours phoning corner drugstores to inquire whether they had Prince Arthur in a can and, when they said yes, asked if they would let him out), you likely experienced the social network phenomenon that we call the *Cyrano Syndrome*. As in Edmond Rostand's classic play, the physiognomically challenged but brilliant poet Cyrano de Bergerac enabled his intellectually challenged but physically compelling friend Christian de Neuvillette to woo and win the beautiful Roxanne. Cyrano aided Christian by sitting behind a bush as Christian stood beneath Roxanne's balcony, intoning words *sotto voce* that Christian could repeat verbatim and thus render an indelibly positive impression on his would-be lover.[6] Ask a question of a teenage clerk; she repeats the question word for word to the store's owner; the owner furnishes an answer; and then she repeats word for word (if you're lucky) what she just heard to you. If only this approach worked as well for small-town retailers as it did for Cyrano and Christian (at least, during the seduction phase)! It may not, but it's a classic example of localized connectivity, without aid of technology, that's practiced every day in small businesses everywhere.

Service on a Limited Scale

As service businesses expand their scale and scope of operations, it's nearly impossible to find front offices that are devoid of information processing or automation—machines often do a better job playing Cyrano to human labor on the front line than people do—but there are corporations of limited scale that still rely largely on people enabling people to get front-office jobs done. Consider Nordstrom, the department store operator that originated in Seattle and has since achieved its apotheosis as an upscale retail industry icon. Nordstrom targets up-market consumers of fashion-forward apparel and home furnishings with a selection of high-quality goods, reasonable if elevated prices, and, most important, personal and attentive customer service. By all accounts, Nordstrom has established the gold standard for outstanding service in retail, setting it apart from other high-end retailers with excellence that's spawned many an urban legend.

Service quality gurus relish the stories about customers' fateful encounters with the store. There is the tale of a woman who came to Nordstrom to return apparel she bought at another department store only to find the items graciously accepted for credit. There is the tale of a customer looking for a suit at Nordstrom and coming up empty-handed; a Nordstrom salesperson escorted him across the street to a competitor, found the right suit, bought it there, and then resold it to the customer on Nordstrom's sales floor. There is the tale of the man who purchased a complete set of snow tires from a tire shop, found them unsatisfactory, and went to Nordstrom to return them; though Nordstrom did not sell tires, the store accepted the return and willingly gave him credit. Because those examples are extraordinary, not to mention ostensibly noneconomic, it's easy to regard such legends with skepticism. After all, if most retailers operated in this way, they would rapidly go out of business. To make sense of Nordstrom's behaviors, you have to know that the store committed long ago to a retail strategy predicated on maximizing the lifetime value of its most loyal customers. This means that not everyone walking in off the street with a set of snow tires can so easily realize store credit. These liberal policies work only if people or systems can identify and track both prospects and customers who fit the loyalty profile; only they are worth the investment of otherwise outrageous acts of service. Since Nordstrom has long operated without sophisticated customer-tracking systems or loyalty programs, it must generate similar results with an interface system composed of people enabling people. That requires in-store organizations oriented around a common understanding of the store's target customers and how it maximizes its return on interactions and relationships with them over time.

Nordstrom's personnel mirror the chain's customers—well educated, reasonably affluent, and stylish in their self-presentation. Its salespeople present themselves in ways that physically resemble the store's customers; they have or project the impression of similar intellectual resources; and they behave with an attitude reflecting polished professionalism. This facilitates their taking a consultative or relationship-based, as opposed to a sales-oriented, approach to the store's customers. Because of their sophistication, these salespeople can differentiate among loyal customers (many of whom they likely know), prospects (who resemble the loyal population but are unknown to them), and mere visitors (who hold less commercial promise). Nordstrom's salespeople do extraordinary things for those prospects and customers who, they believe, will result in extraordinary *returns on relationship* over time. The store's people-dominant interface system still requires information-

processing functions, along with the projection of physical, cognitive, and emotional attributes, but the system's work is largely
done inside the heads of personnel and the social networks they
create within the stores. Nordstrom recently undertook experiments with handheld devices to automate certain cognitive tasks
such as customer tracking and profiling by salespeople on its retail
floors. But the stores remain, in essence, people-dominant interface systems.[7]

You might draw similar conclusions from another exemplary
service organization, Four Seasons Hotels and Resorts. One of
the world's premier hospitality operators, Four Seasons has from
its beginning built a differentiation strategy based on people-
dominant interface systems. The chain demonstrates what well-
managed professionals can achieve in delivering world-class service
through carefully specified human systems. Like fashion apparel at
Nordstrom or gaming in Monte Carlo, Four Seasons accommo-
dations are hardly positioned as bargains. Rooms and meals are
often exceedingly expensive. For certain segments of business
travelers and elite individuals, however, they represent great value.
According to recent travel polls, the rationale has consistently
been "outstanding service."[8] While the chain has many attractive
properties, its focus historically has been, metaphorically speaking, on the "software" of hospitality (people and organization)
rather than the "hardware" (real estate, architecture, and IT).
Four Seasons has consciously chosen not to pursue guest-tracking
programs, loyalty cards, or customer profiling, despite moves by
others in the industry to invest heavily in such systems. It has focused largely on people supporting people in serving guests.
Isadore Sharp, the company's founder and CEO, has commented
that the primary driver of the chain's success is the Golden Rule
("Do unto others, as you would have them do unto you"). "It isn't
about the amenities in the bathroom; it isn't about architecture; it's
all about people dealing [with people] in a sincere way," Sharp explained in a *Forbes* magazine interview. "We decided early on that
if we could have people who could deliver a consistent quality of

service that was better than [our competitors'], that would be our ticket to being the best."[9] To realize this vision of hospitality, Sharp has remained focused on properties that enable the software of people and organization to do its work, operating only hotels at a scale of two hundred to three hundred rooms, where success is predicated on being able "to go anywhere in the world and turn a group of seemingly ordinary people into an extraordinary work force."[10]

Like Nordstrom, Sharp has built operations of modest size across a chain of limited scale—just fifty-seven hotels in twenty-seven countries. Like the casino in Monte Carlo, if either Nordstrom or Four Seasons had opted for larger properties and more intensive distribution, it might have found it difficult to maintain service quality without investments in front-office automation. (In contrast, consider Marriott, which is the largest hospitality operator in the world; it manages over 2,800 hotels with more than 500,000 rooms worldwide and has, out of necessity, invested heavily in information systems, currently operating a loyalty program with over 20 million card holders.[11]) To assure the professionalism of their physical presence, Four Seasons puts all its personnel in uniforms. Four Seasons relies on the cognitive and emotional talents of its frontline staff and their social connectivity with one another to keep track of loyal customers, reward them for their loyalty, and recognize them consistently with personalized service. On a limited scale, human interface systems can pull this off, but it requires that everyone in the organization be aligned with a clear and common set of priorities, operating rules, and values.

For example, New York City's "21" is famous for its black-tie *maîtres d'hôtel* who infallibly remember every guest after a single visit, who seat guests for lunch according to their social and economic prestige (and the strength of their historical relationship with the club), and who guide a cadre of waiters and bartenders to ensure that "21" always has the right information to anticipate guests' needs. This often starts with the establishment placing a returning customer's favorite drink at the table before the guest even sits down. While there is enormous art involved in operating a restaurant at such an elevated level of service, getting the operation right at "21" depends disproportionately on the simplicity of a single location and the tight focus on a small cadre of loyal and targeted patrons. In a business of this size, hiring small numbers of people according to desired physical, cognitive, and emotional attributes, and ensuring that they operate within appropriate social networks to facilitate efficient information transfer, is feasible, if not easy. It's the same formula that has placed the Hotel Okura in Tokyo in the top ranks of places to stay for business travelers in the world. The Okura is widely admired for its service quality. However, if you were to judge the hotel from its real estate or information systems you would be sorely disappointed. Its architecture, inside and out, is a picture-perfect time capsule of the 1950s heyday of the lounge lizards, when Frank Sinatra and the Rat Pack dominated popular culture. There is little front-office automation anywhere to be seen. The hotel's location near the American Embassy makes it a favorite of U.S. travelers, but that's not what has defined its greatness; its greatness lies in the depth and quality of interactions that the hotel can deliver to its loyal guests—the software, not the hardware, of what it does—involving impeccable attention to customer needs, recognition of returning guests, and extraordinarily subtle but polished hospitality.

Four Seasons, of course, involves real organizational complexity, operating on a global stage and across dozens of properties. Its ability to deliver outstanding customer interactions starts with recruiting the right people into the front lines. As Sharp says, "It's easy to train someone to do a job, but it's very hard to train somebody with a poor attitude to be highly motivated."[12] That policy addresses the job's emotional or affective prerequisites. The chain interviews every prospective employee four or five times, evaluating the cognitive capabilities as well; each hotel's general manager conducts one of those interviews. For new hires, there is a seven-part training program with a designated peer trainer operating alongside each new employee, demonstrating what Four Seasons considers the right way to do things, including such details as how to clear a table in a restaurant after a meal. This kind of rigor carries over to job descriptions, levels in the organization, and promotion.[13] You might argue that Four Seasons pays such intense attention to its employees' human qualities precisely because it has few enabling or control systems to routinize the performance of its interface system other than its people themselves. This is the challenge of delivering service at greater than local scale in worlds devoid of machines. Satisfying customers consistently and building loyalty over time across a chain requires that delivery of services take place in consistent ways. At the same time, because high-end service interactions are social interactions, mechanized approaches seldom generate satisfied customers. That's why Four Seasons has emphasized the values of professional rigor and discipline as well as individual latitude and initiative. In the pursuit of excellence, many hotel chains have trained their frontline employees to act with machinelike consistency in word and deed. You might consider this the bitter fruit of Frederick Winslow Taylor's industrial innovations, wherein the time-and-motion studies that were conceived to replicate certain optimal behaviors in a factory (the best way to install a rivet or tighten a bolt) have become the operating logic for behavioral training in settings like hotels. When Taylorism meets the social aspects of services, it can have eerie effects, like the 7-Eleven clerk at the register who says "Have a nice day!" without making eye contact, the fast-food clerk who asks "Do you want a beverage with that?" with emphatic indifference, or the airline personnel who say "Thank you! Bye-bye!" like programmed automatons as you disembark.

Consider the Ritz-Carlton Hotels. Operated by Marriott, Ritz-Carlton has taken a dramatically different approach from that of Four Seasons to operating luxury hotels. Each employee carries a card with the chain's philosophy and operating principles, which begin with the admonition, "We are Ladies and Gentlemen serving Ladies and Gentlemen."[14] Every Ritz-Carlton employee is instructed to greet guests in specified language, such as, "Good morning, Sir" or "Good evening, Madam," and to do so when they are within a certain predetermined distance while walking down a hallway or across the lobby. They are also instructed to respond to guests' requests in precisely scripted fashion. For example, if a guest asks directions, employees must take her there, not simply point the way; if a guest asks for something, the employee should not say "Yes," but "My pleasure." If you are talking to an operator and need to reach another guestroom, she won't say, "I'll connect you," but rather, "Please allow me to extend your call." If you call room service, the staff person who answers will not say, "How might I help you, Sir?" but rather, "How may I serve you today, Mr. So-and-so?" It all sounds terribly civilized until you wake up one day after too many Ritz-Carlton interactions and realize that you're dealing with fellow human beings who are unwittingly engineered to mimic members of Edwardian drawing-room society. Of course, with limited exposure, this kind of strategy for controlling human service interfaces does work, and many guests appreciate how Ritz-Carlton's "Ladies and Gentlemen" relate to them; on too many repetitions, however, the flavor of Taylorism tends to undermine the essential humanity of frontline personnel, resulting in the irony that an inanimate interface, such as a Web site or a voice-response unit, can suddenly feel more human in a service interaction than an actual human being.

At Four Seasons, despite the many rigidly dictated procedures and dress code, the chain eschews strict and programmed behaviors among its employees. Four Seasons instructs its staff to avoid repetitive or stilted phrases, such as "My pleasure" and "I'll be happy to." The result, you might argue, is the truly human use of human beings in a world-class service operation, where individual personality and creativity is not suppressed by the requirements of an interface system. Four Seasons furnishes a profound example of the difference between industrializing services and delivering services at industrial scale. As we will see, there are often limits to the scale and scope of service operations that rely on interface systems composed solely of people. Paradoxically, it's technology that enables large-scale service systems, like Harrah's, to deliver personalized interactions with mass-market reach. The absence of technology often requires managers to utilize rigid control systems governing human behaviors on the front lines, which is why we argue that machines, ultimately, will allow people to be more human in their jobs, not less. But we run the risk of getting ahead of our story, because there are a few businesses that can, and do, manage large-scale operations with people-dominant interface strategies.

Service at Scale

Among service businesses operating at scale without high levels of dependence on front-office technology, few are as impressive as the point-to-point, no-frills, discount air carriers, such as Southwest Airlines and its imitators: JetBlue Airways and AirTran in the United States and Ryanair, Virgin Express, and easyJet in the United Kingdom and Europe. While JetBlue and easyJet, in particular, have deployed technology in the front office to drive efficiencies (as noted, Southwest deployed a self-serve ticketing terminal called the LUV Machine decades ago, but more recently JetBlue has pioneered the paperless cockpit and installed entertainment systems with DIRECTV service in its planes, and easyJet sells tickets online that have no form other than confirmation numbers presented at flight time[15]), the entire class of no-frill carriers competes largely on the basis of its frontline service workers for differentiation of its low-cost, low-price offerings. Without much reliance on front-office automation, these companies have found ways to deliver consistently satisfying interactions for customers despite large work forces, multisite operations, and challenging operating conditions. The low-cost carriers have achieved this level of customer satisfaction by taking integrated approaches to human resources, operations, and marketing in an operating system that represents a striking contrast to that of the full-service carriers. In so doing, this new breed of airlines (except Southwest, which began operating in 1973) have arguably become the most successful airlines in the world. Southwest alone has a market capitalization that exceeds that of all other publicly traded U.S. airlines combined. Ryanair is worth more than eight times as much as Continental, and JetBlue is worth nearly twenty-two times as much as US Airways.[16]

There are many reasons why the low-cost carriers have achieved such success. First and foremost is the simplicity of their operations. Because their focus is on short-haul segments, they fly one type of route—the city-pair—with an average flight time of just over an hour. Since flights are short and don't connect with other airlines, these carriers don't need to operate out of major airports or use hub-and-spoke route systems. With only short-haul flights, they are able to operate one type of aircraft (Southwest and Ryanair fly Boeing 737s, JetBlue flies Airbus A320s[17]), which dramatically simplifies every aspect of maintenance, parts, and training across their operations. Since flights are short, they don't generally assign seats, print boarding passes, serve meals (other than Southwest Chairman Herb Kelleher's favorite airline meal, "filet of peanut," and snacks for purchase on the other carriers), operate multiple classes of service, or accept interline baggage.[18] After all, who needs food or your favorite seat when you're only on board for a short hop? And how many people who take a day trip need to check a bag? The result is that Southwest and its brethren can "turn" their aircraft in an average of fifteen to twenty minutes, as opposed to the industry average of fifty-five minutes. (When airlines turn aircraft, they disembark all passengers on arrival at the gate, clean and provision the aircraft, load new passengers on the plane, and push off from the gate.) When an airline serves everyone on board multiple-course meals, cleaning takes time, and so does catering. When passengers are assigned seats, boarding takes time, since airline staff need to pay more attention to boarding passes and resolve seating disputes. Each of those versions of simplifying airline processes enables these airlines to lower their turn times. The leverage in savings that low-cost carriers realize in turning planes is enormous, since they fly their aircraft an average of ten segments a day as compared to five at full-service carriers. Indeed, easyJet's founder, Stelios Haji-Ioannou, claims that turn-time efficiencies at his airline mean that easyJet can use two planes to do the work of three and thus operate a smaller fleet relative to competition.[19] The economic advantages of this operating model are easily seen in the relative cost levels across airlines. As the highest-cost carrier in the United States, US Airways incurred costs of 15.8 cents to fly one air-passenger seat mile in 2003 (the industry's standard cost metric), while Southwest incurred costs of only 7.6 cents and JetBlue of 6.1 cents.[20] No wonder the major airlines can't compete with this new breed of airline on price!

Every one of the low-cost carriers has adopted a people-focused approach to differentiation of services: each of the airlines relies on people-dominant interfaces to make interactions with passengers uniquely satisfying and loyalty inducing. At Southwest, JetBlue, Virgin Express, and Ryanair, the frontline staff has acquired a reputation for friendliness, humility, frugality, efficiency, a sense of humor, and iconoclasm. By simplifying operations to reduce errors and delays, and recruiting frontline service people with an eye toward differentiation, these carriers have created a new kind of air travel experience. In this regard, the way Southwest makes new hires is instructive. In line with Kelleher's dictum, "Hire for attitude, train for skill," the airline has developed creative approaches to ensuring that it puts people with the right emotional or attitudinal attributes on its payroll.[21] For example, Southwest invites job applicants to "audition" for their jobs in large hotel ballrooms. With a hundred or more people present, a Southwest employee will ask candidates to stand up and tell a story about the most embarrassing moment in their lives. What might appear like a recipe for public humiliation is in fact an effective screening process. The company's philosophy dictates that working at Southwest should be fun; this is the airline that originated in Love Field in Dallas and advertised in the 1970s as "The Airline That Luv Built."

(Southwest's ticker symbol on the New York Stock Exchange is LUV.) At Southwest's job auditions, the decision makers are airline employees and frequent flyers, based on a logic that says no people are better positioned to evaluate a potential hire than the people whom they will serve and the people who will work alongside them. Telling personally compromising tales before a large group of potential colleagues is a well-engineered test for critical service interface attributes that Southwest people require, such as a sense of humor (jokes make repetitive tasks more palatable); an ability not to take oneself too seriously (airline work is often menial); the capacity to command attention from crowds (Southwest's 737s seat 122 to 137 passengers); and a personality that's fun to be around (this promotes teamwork). The result is an airline with more than thirty-five thousand employees that still operates, in the words of Colleen Barrett, Southwest's chief operating officer and corporate secretary, "like one big happy family." The family ethic has practical implications: Even with a work force that's 81 percent unionized, Southwest's employees consider their jobs flexible, and everyone pitches in. When turn times run long, pilots descend to the tarmac to help load bags.[22]

The benefits of this human resource strategy are airlines with differentiated interface systems composed largely of people. While low costs make short-haul carriers economically viable, the culture their people instill in flight and at the airports makes them distinctive. At JetBlue, passengers remark constantly about the hard-working and dedicated attitude of employees at the airline, who seem to enjoy their jobs and excel at communicating intelligently with customers. While their pay rates are somewhat lower than airline industry averages (in line with other discount carriers), JetBlue's so-called crewmembers benefit from a positive and motivating work environment, a generous employee stock purchase program, attractive opportunities for rapid career advancement, and a profit-sharing plan in which 15 percent of the airline's net income is distributed to its employees. It does not hurt that David Neeleman, JetBlue's CEO, has jumped into the trenches in times of crisis. During the blackout in the New York area in 2003, Neeleman himself showed up to tend the stranded passengers at John F. Kennedy International Airport, where JetBlue quickly distributed food, $25 airline credits, and ultimately became the only airline that day to resume operations in spite of the power failure.[23]

At Southwest, the corporate headquarters is filled with pictures of employees and their families, and people throughout the company hug one another and offer boisterous greetings in Kelleher-like fashion. On the aircraft, Southwest employees are known for their pranks—which Ryanair and Virgin Express, in particular, have emulated—such as hiding in overhead bins to surprise passengers getting on the plane, wearing costumes for holidays, and running contests focused on a wide array of personal peccadilloes among passengers. For some, Southwest's humor and frivolity are repellant. After all, if you want an assigned seat, a balanced meal, an airline connection, and a checked bag, Southwest and Ryanair are not for you. Indeed, you're likely to experience interactions with these airlines as truly objectionable service experiences. But for target customers, the frat-house-in-the-air atmosphere suits them just fine. It is the human face of these airlines—combined with the underlying reliability of their operations—that endears them to frequent flyers. After all, low price is seldom the basis for an emotional bond, but intelligence, affect, and community are. The low-cost carriers start by creating effective and compelling social networks within their people-dominant interface systems (populated by individuals who have the cognitive and attitudinal attributes to perform successfully with teammates in their jobs), and then they invite customers in. In this sense, the social interaction of staff becomes the presentation layer for the airline's transportation assets and services.

Consider the contrast between Southwest and US Airways when it comes to a detail as basic as the FAA-mandated safety demonstration before takeoff. From one perspective, US Airways is far more sophisticated. Rather than having its in-flight personnel display a physical seat belt and oxygen mask to give the demonstration, on most of its planes scores of flat-panel displays descend from the ceiling panels to play a safety video and retract when it's done. Is it efficient? Absolutely! Is it effective? That's a more difficult question to answer. Most passengers tune out anything that looks remotely like television; and most flight personnel are not doing much else just before takeoff, so the productivity gains are debatable. Still, videos get the job done. At Southwest, giving the safety briefing is the highlight of the flight—for cabin staff and for passengers. Southwest employees relish these opportunities to perform and find ever more creative ways to make the otherwise rote and repetitive information engaging. The presentation is a crowd pleaser for frontline employees and customers alike, and it gets the important messages across to jaded travelers by attempting to engage everyone.

Of course, the short-haul carriers put tens of millions of lives at risk every week, and hence the integrity of their operations is no laughing matter. Arguably, they expose themselves to more risk because of the high frequency of takeoffs and landings. Yet, these are among the safest airlines in the world. In its three decades of operation, Southwest has never incurred a fatality.[24]

How is this possible? There are several answers: Reduced complexity of operations results in less likelihood of errors. Positive employee morale results in fewer mistakes. Satisfied customers are more tolerant (in line with research we discussed in chapter 3) and create fewer problems. Loyal customers work with, not against, loyal employees in making the system work. These software attributes reinforce the hardware advantages of simplified route systems, less congested secondary airports, fair-weather destinations, newer planes (JetBlue's fleet has an average age of 15.5 months as compared with Continental's at over seven years[25]), and, as noted, fleets composed of one kind of plane. The front-office operations of these airlines work, however, based on interface systems in which everyone shares a common set of human values, understands how the airline works, and how it succeeds as a business. This common understanding enables frontline workers to run the airline day in and day out with energy left over to interact with customers as human beings. The result across the low-cost carriers is the emergence of several highly distinctive brands in an industry where the operating elements (planes, airports, ticket counters) are utter commodities. This is why the low-cost carriers have proven so difficult for the majors to imitate, despite Southwest's example operating in their markets for decades. Delta's Song has hired designer Kate Spade to give planes and uniforms (the hardware) a new look, but the people and organization (the software) remain largely the same. While Song has changed many of the logistical aspects of operations around employee hours per month, plane fleets, and turnaround times, even reducing operating costs vis-à-vis Delta, early data suggests that Song is stealing share from other Delta units, rather than other low-cost carriers like JetBlue.[26] If, as Delta officials claim, Song is on equal footing with JetBlue in terms of logistics and cost, then the key difference between the two lies in the employees themselves and the interactions they facilitate.

All of which simply highlights the competitive advantage intrinsic to successful interface systems, especially those composed largely of people. The dynamic interplay of humans in an interface system is easy to observe but difficult to replicate; and it's particularly challenging to operate at scale. The cultural attributes of such people-dominant systems may take years to nurture and grow. This is why technology often becomes an essential driver of any corporate change process that aims to rapidly alter attitudes and behaviors in interactions on the front line.

ATTRIBUTES OF PEOPLE-DOMINANT INTERFACES

Each of the businesses we've visited in this chapter puts different emphases across the four attributes of interfaces within their interface systems, as dictated by their competitive strategy. At Four Seasons, for example, the appearance of frontline staff—uniformed, clean-cut, businesslike, courteous, individual, and authentic—enables differentiation of interactions on a physical dimension. At

Nordstrom, the average salesperson's ability to recognize and reward the store's best customers with appropriate service and attention enables differentiated interactions on a cognitive dimension. At Southwest, Virgin Express, or Ryanair, fun and iconoclasm among flight attendants create interactions with passengers that are differentiated on an emotional dimension. And the free-flowing communications that support a Four Seasons staff in orchestrating a seamless hospitality experience at each property, enable Nordstrom salespeople to transfer customers gracefully from one department to the next, and make it possible for easyJet's crews to work as a team in flight are a function of social network or community connections within organizations. We could break the success factors of each organization down according to each of the four attributes and illustrate where trade-offs have been made. (For example, using humor or affect to manage customers' anxiety is more critical in air travel than it is in hospitality or retail.) The point, however, is clear: These four attributes represent the dimensions of meaningful interactions and relationships between companies and their customers, whether mediated by people or machines.

WHEN PEOPLE-DOMINANT INTERFACES GO WRONG

For every success story among service businesses, there are many more tales of failure and dissolution. When technology plays no role in guiding human behavior on the front lines, success depends, as we've seen, on shared culture and values, which are rare commodities in any business organization.

Consider Sears. In the 1990s, after divesting itself of its pioneering catalog operations and the Discover Card business, Sears chose to focus on delivering a differentiated customer experience in its brick-and-mortar stores. Despite a much-publicized turnaround in the late 1990s, which included articles about Sears's employee-customer profit chain model for managing retail services, the franchise continues to struggle. One problem: The orientation of Sears's people could be characterized as low on every one of the four attributes we've discussed. For example, when Sears managers attempted to assess the cognitive and emotional grasp of the business by the company's retail work force, which then totaled 275,000 frontline employees, they got stunning data in return. When asked, the primary job-related goal stated by more than half the employees on retail floors was "to protect the assets of the company."[27] Their response was both wrongheaded (retailers display goods in big-box formats so customers can interact with the merchandise) and wrong hearted (retail work forces don't succeed when they view the customer as the enemy). Perhaps service-oriented and customer-centric Lands' End, which Sears acquired in June 2002, can change the souring behemoth's culture. But an acquisition of a best-in-class direct-mail retailer is unlikely to prove sufficient to save the brand. Reengineering the front office is what's required, and service in a mass-market setting is unlikely to succeed through people alone, especially when the task requires a cultural reawakening in a population that's the equivalent of a small city but also geographically dispersed.

Sears is not the only big-box retailer dealing with human resource challenges in the front office. Enter any large-format retail franchise, and you're likely to have an indifferent experience. Home Depot, once famous for its skilled salespeople, cannot find enough former contractors to lend customers a hand on its warehouse floors.[28] Office Depot, once known for its skill in meeting the needs of small businesspeople, has eleven hundred stores in North America with twenty-three thousand employees who offer undifferentiated service to shoppers. According to one recent advertising agency survey, a minority of shoppers in the Big Three office-supplies superstores had any idea, *while they were in the store*, whether they were at an Office Depot, OfficeMax, or Staples.[29] Toys "R" Us, once the category-killer for children's products, now offers impersonal service by employees who cannot locate SKUs, let alone remember what it was like to be a child; as a result, the chain now plans to exit the toy business![30] And such experiences are hardly limited to retail settings.

There is a crisis in business resulting from a dearth of the right people, with the appropriate physical, cognitive, emotional, and social attributes, to staff the front lines. The vast majority of companies today cannot manage meaningful interactions with their customers by relying on the people they currently employ. Whether their employees lack the basic cognitive faculties, don't have empathy for customers, or cannot work successfully in teams, people-based operating strategies for competing on quality of interactions are imperiled.[31] Not only are customers in short supply, but so are the skilled individuals who might serve them and meet their needs. That is an ironic state of affairs at a period in business history when more companies than ever find themselves competing on the quality of their sales, service, and customer care. And more workers than ever are looking for jobs. This context is where front-office application of technology in the interface system can make a difference.

Conclusion

The paucity of companies that achieve breakthrough performance in managing customer interactions and relationships is testament to the challenges of getting the traditional front office right. At small limited scale, people supporting people may constitute a robust interface system, but there are precious few smaller service providers who can claim even to be merely good. Those that transcend the merely good, like Nordstrom or Four Seasons, become legends in their own time. Even In-N-Out Burger has been feeling the pain as it has grown, now deploying employees armed with wireless PDAs to take orders and prevent lengthy lines at the pick-up window. Especially among operations of significant scale, outstanding interface systems are the exception, not the rule. There are several insights worth bearing in mind.

- *Human and machine interfaces share critical attributes that render them capable of mediating interactions and relationships between companies and customers.* These attributes correspond to the four drivers of technology evolution—physical presence and presentation (proliferation), cognition (intelligence and inter-activity), emotion or attitude (affect), and social networks and interpersonal skills (ubiquitous connectivity). These dimensions of interface capability in people correspond to those of machines, because we are, in effect, sourcing machines to play frontline roles. The people-dominant interface system is the approach to service management that technology enables us to reengineer.

- *To deploy people-dominant interface systems at scale is challenging because affordable and appropriately skilled people are hard to find.* Competing based on people-dominant interface systems is challenging in competitive contexts that demand consistently excellent service delivery. That's why innovative strategies for managing interface systems, involving people and machines, are becoming an imperative across many industry sectors. For companies deploying people-dominant interface systems alone, the challenges are myriad, especially with respect to human resources. Since frontline workers must excel without the benefits of machine or systems support, companies must compensate with greater investments in training and skills development. Even so, many businesses today simply entail too much complexity for people, unaided by technology, to master the tasks required.

- *Front-office automation is rapidly becoming an imperative for both productivity gains and growth.* It helps if back-office operations run flawlessly like Southwest; otherwise, frontline personnel cannot focus on managing customer interactions and relation-ships. Even so, there are inevitable limits to scale.

- *As businesses become more complex, humans alone can no longer perform many frontline service jobs—machines in the work force are for most companies a necessary condition of success.* In this chapter, we have focused on a series of businesses that were able to succeed using people-dominant service interfaces. But those are businesses that across the board do not challenge average frontline workers to deal with voluminous quantities of dynamic or complex information intrinsic to service delivery. Many businesses have become increasingly complex, and the cognitive requirements of frontline work have exceeded the grasp of human beings. In chapter 5, we turn to businesses where machines operate as the primary service interfaces between companies and customers. Such interfaces are another basic building block of strategy and operations in the reengineered front office—and the second of our interface archetypes—as companies work to respond successfully to increasingly demanding customers.

5

WHAT MACHINES
DO BEST

IN OUR DISCUSSION OF FRONT-OFFICE REENGINEERING thus far, we've explicitly framed the fundamental question of how companies can best manage interactions and relationships with customers: What do people do best and what do machines do best? After all, people and machines each have strengths and weaknesses. While you might always prefer to have the best people relating to customers, those people are not available everywhere at once—and, if they were, they would reject the pay scales viable in, say, mass-market retailing. Managers at Sears or Home Depot cannot count on access to hundreds of thousands of experts on their front lines. Instead, those chains must plan with more realistic expectations about whom they can hire to manage customer interactions in their stores. To compensate for their employees' lack of expertise or knowledge, such businesses must use capital to assist labor (machines supporting people) or to substitute for labor (machines replacing people); and they must choose between performing such services proximally (in stores) or remotely (through network connections to capabilities off-site). The alternative—namely the assured degradation of customer interactions and relationships in the stores—is clearly unappealing and ultimately noneconomic. In this chapter, we examine the second of our three interface archetypes, the substitution of machines for labor. The growing evidence from consumer markets, in particular, suggests that machines can perform ably on all four of our interface attributes—physical, cognitive, emotional, and, if you will, synaptic—to become compelling relationship managers on their own.

Consider the resurgent Apple Computer. Apple's new corporate stores are extremely appealing retail environments, superb showcases for fostering higher levels of awareness, connection, and even intimacy between its customers and its products. The retail stores blend people-dominant interfaces (product-savvy store clerks) and machine-dominant interfaces (Apple products on display, plus large flat-panel monitors that run promotional videos to entertain and educate visitors). The stores, however, exist primarily to provide a context in retail space that maximizes the aesthetic and ergonomic appeal of Apple's products—which are Apple's primary relationship management interfaces with its millions of customers. For Apple, the product is the best ambassador of the brand and the best service provider for the customer. As such, each product is a machine-dominant service interface that scores high on physical attributes (Apple's machines are aesthetically striking and distinctive), as well as cognitive (the Power Mac G5 had the first 64-bit processor), emotional (the Macs and iPods elicit strongly affective responses among users), and synaptic (Apple's products are highly interoperable with networks for accessing movies, music, other content, other machines, and, most important, other people).

What people-dominant interfaces do for Southwest Airlines and Four Seasons Hotels and Resorts, consumer products do for Apple. One could argue that, for many companies, the products are the faces and voices of the brands, whether those products are consumer durables such as cars and personal computers or packaged goods such as perfumes and DVDs. Products interact with users more often than frontline service workers or advertising campaigns do. One might conclude that every product is a service interface. If a product succeeds, it strongly determines customers' perception of value over time, and that, in turn, increases or decreases customer satisfaction and intent to repurchase. Of course, some products are more compelling than others, just as some engage customers more intensely than others. Consider how popular Web sites manage relationships with customers. Regardless of the televised advertising campaigns of eBay and Yahoo!, their Web sites are their brands' primary relationship managers. Their sites are also forms of media that provide services. In this sense, both products and Web sites often function as machine-dominant service interfaces—backed by people. This chapter focuses on how the machine, one that mediates customer interactions (assuming only today's processing capacities), might affect businesses that seek to substitute capital for frontline labor.

FRONT-OFFICE MACHINES OF THE PRESENT

From an historical perspective, machines and products have explicitly been providing services for a long while. Books substituted for storytellers, broadsheets for town criers, slot machines for card dealers, vending machines for soda jerks, juke boxes for disc jockeys, automatic lifts for elevator operators, ATMs for bank tellers, and so on. Scores of mundane customer-facing activities, once performed by people, are now performed by machines.

The results of such machine-based changes surround us at home. Bread machines act as bakers and coffee machines grind and brew the beans; caller ID and voice mail screen calls and take our phone messages; and alarm systems and smoke detectors safeguard our houses. One need only visit less developed countries to see how different daily life can be without these machine-based innovations.

Outside the home, capital is laboring everywhere. Hertz Rental Car customers once relied on frontline staff for driving directions. But Hertz rents 30 million cars a year; it couldn't possibly train the available work force to give good directions, even if workers at each rental location could handle the task.[1] So Hertz put automated kiosks in every rental office to print directions on demand. In 1995, Hertz got even closer to its customers by installing NeverLost systems in many of its cars.[2] NeverLost uses global-positioning satellites to locate vehicles on street grids and then directs drivers, with verbal cues, to their destinations. Unlike the kiosk service, NeverLost guides customers throughout their rental experience through a naturally appealing interface—a (synthesized) human voice. While kiosks are efficient to use, NeverLost (despite its relatively primitive technology by today's standards) exerts a uniquely emotional appeal on drivers, starting with its brand name, which promises to keep customers safe and on course.

Machines perform many of these tasks even better than the most skilled workers money can buy. For this reason, we sometimes prefer machine service over people. Dishwashers wash dishes better than we do; and, even with its flaws, NeverLost outperforms maps used by hapless travelers in unfamiliar cities. Automated teller machines dispense cash more quickly and accurately than bank tellers, and Fast Lane and E-ZPass move us through tollgates faster than human toll collectors. Driverless monorails operate more reliably and consistently in airports and theme parks, and automated vacuum cleaners and lawn mowers take no cigarette breaks or sick days. Airport e-ticketing machines check in passengers faster, and ATM-pumps at gas stations enable customers to bypass indifferent or unsavory clerks.

In the travel sector, the trend is perhaps most pronounced. Hertz's Gold Aisle allows gold-card members to pick up cars without a service representative's assistance. National Car Rental lets loyalty-card–holders select *any* car from a special lot where they can sit in the vehicles before choosing their car and can exit by swiping their membership card at the gate—without ever interacting with human staff. The hotel chain Club Quarters enables members to check themselves in and out through an ATM-like apparatus in the hotel lobby; these machines generate room keys and billing statements. Hilton installed kiosks that allow guests not only to perform similar functions but also to review bills, get directions to rooms, and, through partnerships with airlines, print boarding passes.[3] Hyatt is working on a system that would allow guests to check in and out using cell phones and payphone-sized devices in its lobbies. So mainstream are these innovations that the *Wall Street Journal* ran a headline, "How to Have a Pleasant Trip: Eliminate All Human Contact."[4]

FRONT OFFICE OF THE FUTURE

The business world has swiftly replaced many *face-to-face* with *screen-to-face* interactions with customers. So why do doormen persist, even though most no longer open doors? A skilled doorman welcomes, dignifies, and reassures people; his humanity transcends opening doors so that guests remember interacting with him. Until now, machines on the front lines of business have played only blue-collar roles—efficiently dispensing cash, opening doors, providing transportation, dealing cards, cleaning floors.[5] Now their *raison d'être* must extend beyond efficiency: they, too, must transcend the underlying task.

For example, refilling prescriptions through phone-based automated systems reduces delays and increases accuracy—a straightforward transaction. Growth in the category forces pharmacies to pursue automated service solutions: prescriptions are expected to rise at least 26 percent between 2001 and 2005, while the number of available pharmacists in the work force will increase by less than 4 percent.[6] If those machine-based solutions further commoditize drugstore offerings, then differentiation among the chains diminishes even as efficiency of operations increases. Pharmacies have, therefore, two options: Freed by machines from doing rote work, pharmacists can become more effective relationship managers just as family doctors once did on house calls; or machines must interact more distinctively and compellingly with customers.

We are only now realizing the promise and potential of white-collar machines. While front-office automation of operational tasks will proceed, machines that make interactions more efficient (addressing cost) *and* more effective (addressing differentiation)

will drive future advantage. Until now, machines were not sufficiently evolved to assume white-collar positions. As computer scientist Hans Moravec writes:

> The first electronic computers in the 1950s did the work of
> thousands of clerks, seeming to transcend humans, let alone
> other machines. Yet the first reasoning and game-playing pro-
> grams on those computers were a match merely for single
> human beginners, and each only in a single narrow task. And,
> in the 1960s, computer-linked cameras and mechanical arms
> took hours to unreliably find and move a few white blocks on a
> black tabletop, much worse than a toddler. A modest robot in-
> dustry did appear, but consisted only of arms and vehicles fol-
> lowing predetermined trajectories. The situation did not
> improve substantially for decades, and disheartened waves of
> robotics devotees. . . . But things are changing.[7]

Moravec concedes that PCs today cannot yet outperform the
human brain and they lag in processing power by several orders of
magnitude. But he estimates that machines will surpass the processing capacity of the human brain in only twenty years. We need
not wait that long to realize the benefits of this evolution. In *The
Age of Spiritual Machines*, entrepreneur Ray Kurzweil describes
machines that are already breaking barriers of artistic expression,
composing original poetry and classical music so persuasively that
informed readers and listeners can not distinguish their output
from that of great composers.[8] In recent years, machines have out-
performed people in complex tasks that supposedly only human
intelligence could accomplish. For example, the community of chess
masters disregarded computerized chess-playing software despite
its increasing prowess throughout the 1970s and 1980s. But when
IBM's Deep Blue defeated the world's reigning chess champion,
Garry Kasparov, in a six-game rematch in May 1997, the world
noticed. Deep Blue's processing power enabled the machine to analyze 200 million board positions per second.[9] Five years later, chess
master Vladimir Kramnik reached a tournament draw of four to four

with IBM's latest machine entrant, Deep Fritz, by playing to the machine's limitations. Deep Fritz had less processing power than Deep Blue but used a so-called pattern recognition pruning algorithm.[10]

Our standing in the race between human and machine intelligence depends on the context in which we deploy smart machines. According to Moravec, we can map out robotic or intelligent-machine evolution in biological terms, thinking through generations of development. Today's lawn-mowing and vacuuming robots (domestic service) belong to what he considers the first generation. They can do simple tasks but cannot adapt to changing circumstances. Within a decade, he predicts, a second-generation robot with the cognitive capacity of a mouse brain will respond to positive and negative reinforcement within predefined circumstances (as Sony's AIBO does today), substantially improving performance of simple tasks over time. By 2040, Moravec expects third-generation robots with monkey-like intelligence, capable of learning very quickly from "mental rehearsals in simulations that model physical, cultural, and psychological factors," endowing such machines with a primitive kind of consciousness and "a simple inner mental life concerned only with concrete situations and people in its work area." Training such machines will require simulation environments to introduce them to everyday situations; the robots would then be prepared to operate in the world and retune to stay "faithful to reality." Such robots would build on the simulators' provision of a "two-way conduit between symbolic descriptions and physical reality." By 2050, Moravec foresees fourth-generation "universal robots" that can "abstract and generalize [and] become intellectually formidable."[11]

Putting Machines in Their Place

We previously set forth four essential attributes of interfaces—apposite to people and machines—that are intrinsic to the performance of interaction or relationship management roles. In chapter 4, we explored what humans do best. In this chapter, we explore what machines do best. In chapter 6, we examine how people and machines can most effectively work together. The idea that people and machines bring different performance capabilities to the labor force is just emerging in the economics literature as an area of inquiry. One researcher has proposed a preliminary catalog of human and machine strengths, where humans excel in "judgment, pattern recognition, exception processing, insight, and creativity" and machines excel at "collecting, storing, transmitting, and routine processing."[12] But this inventory is limited both by today's conceptions of what historical IT investments can deliver in corporate business environments and by current machine capabilities. Clearly, the division of labor between people and machines will evolve as technology advances. But the decisions about what people and machines do will transcend any inventory of performance attributes. We must raise the obvious question: What kinds of service interfaces do customers want? In reverse-engineering a company's appropriate interface capability, we follow a logic chain from the desired customer experience, to the nature of a customer's interactions or relationships with a company, to the service interfaces that enable those interactions and relationships, and finally, to the composition of those interfaces as a system according to an appropriate division of labor between people and machines (see figure 5-1). In this sense, we are choosing among what people and machines should do in the context of a dynamic equilibrium, involving the supply side (the capabilities of people and machines)

FIGURE 5-1

Customer and Company Influence on Interactions

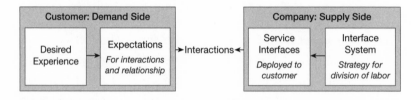

and the demand side (what customers want), and considerations of efficiency (service interactions at lower cost to the company) and effectiveness (service interactions driving higher customer perceptions of value). Today, the effect of machines entering the labor force is culturally, socially, and psychologically complex. Machines are already changing corporate back offices and front offices, white-collar and blue-collar roles. We seek to understand the subtle judgments that will optimize allocations of effort between people and machines across those realms, and so we examine the physical, cognitive, emotional, and synaptic dynamics of human-machine interactions.

The Physical

Machine-dominant interfaces with relationship management capability can take a variety of physical forms, ranging from machines that work with users through push or pull. (We draw from the marketing discipline's concepts of *push* and *pull* communications, where point-of-sale promotion, such as in-store displays and coupons, constitutes push media and mass-market advertising and public relations constitute pull media. One pushes transactions at the point of sale, while the other pulls consumers to the point of sale.) At both ends of the spectrum, we see a veritable explosion in output of large and small devices in numerous designs and incarnations, engineered to interact well with humans. Such *pull technology* as touch-screen kiosks, ATMs, and Web sites, are usually tied to specific locations, often physically connected to networks, and embedded in local geographies (airports, retail stores, bank lobbies, and PCs); they generally perform specialized functions, and we come to them for singular purposes. Consider Shop 2000, the eighteen-foot-wide vending machine unveiled on the sidewalk of the Adams Morgan neighborhood in Washington, D.C. This fully automated drugstore enables users to serve themselves pharmacy items. Described by one wag as "a 7-Eleven in a box," customers have embraced Shop 2000. Many perceive the machine as more accurate and reliable in executing transactions than a typical drugstore clerk—and refreshingly free of attitude. One shopper commented, "A guy in the store can make a mistake or give you a hard time, but not the machine. I definitely prefer the machine."[13] As we have already noted, major retailers such as WHSmith and Staples are experimenting with similar stores-in-a-box.[14]

But pull technology doesn't live with us or follow us around. In contrast, small mobile, portable, or wearable devices such as cell phones and PDAs are what we call *push technology* (not to be confused with previous usage of this term, which referred to companies such as PointCast and Marimba that pushed information over networks). To prove viable, these push devices must appeal to us very differently from pull technology. They cannot merely perform certain services. They must become welcome in our lives and often on our persons. If they are not sufficiently compelling functionally, ergonomically, and aesthetically, we'll leave them behind. If they appeal weakly, we won't interact with them every few minutes, as do users of the BlackBerry (RIM), the iMode phone (NTT DoCoMo), and the Treo (Handspring). Without such interaction, their usage value greatly diminishes. Clearly, pull devices also fail if they're unappealing, but they are not subject to the same high threshold of compatibility and even intimacy with their human users. After all, we don't wear Bloomberg terminals on our belts and we don't carry Fandango kiosks in our purses.

In effect, there are pull machines that consumers tolerate, interacting with them from time to time, and a new breed of push devices with which consumers willingly interact frequently and with which they often form personal attachments. As we have noted, the idea that consumers can fall in love with certain products is not new and not limited to information appliances or interactive interfaces (screen-based, speech-based, or otherwise). The New Beetle from Volkswagen charmed its mostly female drivers with the delightfully quirky personality of the car manifested inside and out, right down to the bud vase on the dashboard. The Beetle embodies impish humor and charm. The Mazda Miata captured the imaginations of its mostly male (and middle-aged) drivers with its 1960s-era soft-top roadster style and engine whine engineered to sound like a sports car from earlier days. The Miata embodies the romance of a classic motoring era, which corresponds to the salad days of its target market, with enough raw performance to make it fun. The visually arresting kitchen gadgets called OXO Good Grips have won over serious home chefs around the world with their strangely bulbous black-rubber handles. Like Apple's iPod and Palm's V, these are products that their owners *want* to use; their appeal goes well beyond functionality—they appeal to users' hearts as well as hands—with or without the interactive capabilities based on smart technologies. Of course, there are exceptions among mass-market success stories, but they are truly isolated cases. Where choice is limited, users may bond with interfaces out of necessity, not affection or attachment. Consider Microsoft's Office suite, with its near-control of the PC desktop and dominant global market share, or IBM's Lotus Notes, where corporate mandates render individual user preferences moot.[15] These software platforms dominate their markets, but not due to the success factors for either push or pull devices.

To understand the contrasting dynamics of push and pull interfaces, look at the automotive industry. You might argue that automobiles are pull technology on the outside and push technology on the inside. Their exteriors are designed to pull customers to them initially and their interiors to push interactions and build relationships over time. Only recently have the Big Three automakers in Detroit awakened to the importance of interior design. In the 1970s when a company like General Motors controlled over half the market share in new car sales in the United States, few managers worried about satisfying customers once they actually bought a car. Though market shares have declined precipitously since then, the oligopolist mind-sets have proven slow to change. Remember Ford Motor Company's ballyhooed relaunch of its retro-design Thunderbird? The new T-bird got rave reviews from the automotive trade press as a concept car at auto shows. But when the car itself hit the market, and reviewers could drive the real thing, they were dismayed; for starters, its interior bore a startling resemblance to that of a Ford Taurus.[16] Unlike the New Beetle or the Miata, which pulled from the outside and pushed from the inside, the stiff and sluggish T-bird turned out to be a humdrum sedan in disguise. By not delivering on its promise, the car failed as a service interface: It failed to satisfy users and induce loyalty where and when it mattered most—while they were driving the car.

In contrast, Asian automakers obsess over how cars *feel* when drivers operate them. For decades, Honda has invested in teams of engineers to improve the feel of interacting with a particular dashboard button or control because Honda understands that the car—not the salesperson on the showroom floor or the advertising on television—is the manager of the relationship with the customer and the ultimate ambassador of the brand. General Motors spent $3.43 billion in advertising and promotion in 2003— the largest marketing budget of any company in the world—and still lost market share.[17] (No wonder analysts predict that one of the Big Three U.S. automakers will disappear in the next ten years.[18])

In evaluating the physical forms of products and machines, we are suggesting that goods of all kinds, not just computers and consumer electronics, must operate as managers of interactions and relationships with customers. In other words, every product is a service interface—partly because of the embedded intelligence of devices and networks, partly because companies must compete more on how than what they sell, and partly because of the extraordinary levels of competition among offerings and brands. One way or another, a product's physical presentation matters— and it matters more than ever. When you can combine that physical presence with cognitive or emotional attributes—as we see in products from the iPod to the iMode phone—powerful bonds form with users. Innovations such as BMW's iDrive and GM's OnStar, for all their initial flaws and limitations, are essentially right-minded; they openly acknowledge that the appropriate role of a complex, expensive machine like a car is to interact meaningfully with customers and build long-term relationships for their brands. In this sense, cars are not only functionally, ergonomically, or aesthetically appealing; their attributes have a social context. That's partly the message of robotics, which involves the development of smart devices in anthropomorphic forms designed to trigger human emotional responses. (Is it a coincidence that in the 1960s TV sitcom, "My Mother the Car," the protagonist's deceased mother continues to rule his life as a voice inside his car radio?[19]) Simply put, because every product or machine is potentially a service interface, its physical attributes are important aspects of its viability.

The Cognitive

To be sure, the world abounds with objects—art, fresh flowers, well-designed tools—with which we fall in love. But little else promotes deeper ties with the things around us than the kind of embedded intelligence that facilitates meaningful or gratifying interaction. That's the lure of the Nintendo GameCube, Sony PlayStation, BlackBerry, and TiVo. Along these lines, can you imagine a more compelling example of cognition built into machine interfaces than the Web? As an enabling platform for compelling interaction with network-based intelligence, the Web puts the cognitive capabilities of some of the world's most advanced machines at the fingertips of mainstream users, who can access the network's vast processing power and database capacity from the interface of their PCs. While the Web may feel like a social or commercial environment, it is actually a front end, or interface, that enables users with limited or no technology skills to access hundreds of thousands of computer servers and databases. Without the Web's interface, through which digital content and online service providers render their presence with personality, attitude, dynamism, persuasion, and humor, the Net could never have become a platform where mass audiences congregated or where mass markets could form. Besides the Web, where else would people sit alone in a room, query databases for hours on end, and call it entertainment?

What has enabled the online commerce revolution? Largely, the data-processing power of online sites. Breakthrough businesses like Amazon operate massive databases that cross-tab deep knowledge of products with deep knowledge of consumers, providing not only the richest product information (commercial, editorial, and peer-to-peer), but also the richest consumer information (what we bought, how we buy, whom we know). The innovations that Amazon tries with users represent quantitatively driven direct-marketing and direct-response programs. The related-product recommendations, the promotional programs, the free shipping offers, and the personalized stores are controlled experiments with stimuli designed to induce consumer purchases. The shopping carts, wish lists, address books, and account information are enabled by database functionality. No matter how personable and friendly the Amazon site may feel, the business represents a cognitive triumph in harnessing machine intelligence to address age-old challenges of mass-market general merchandise and direct-response retailing. A visit to Amazon is like a visit to some idealized version of your local library. The interface of the card catalog, like Amazon's searchable interface, contains information about the products; the library's stacks, like Amazon's warehouses, hold the products. The intelligence of the librarian, who marries his vast knowledge of your queries with volumes in the library's collection, completes the interaction.

Such cognitive attributes are one of the areas where machines excel today. It's not that human beings cannot outthink machines—we mostly still can—but machines have the edge when mastering vast amounts of data and processing it on demand with lightning speed. Indeed, the match-up between a good Web site and even the best retail clerk is no contest. Try finding a clerk in a Barnes & Noble or a Tower Records store who can remember you from one visit to the next, offer recommendations based on your past purchases, introduce you to other customers with relevant opinions, or refer you to neutral sources of category expertise relevant to your purchase. How many clerks can store your credit card information for future purchases, save items in shopping carts for you to buy later, keep track of items you might want to buy, remind you of friends' and family members' birthdays, store your address book, and notify you when items arrive in the store or when orders ship? In such cognitive processes, machines win. The marriage of database-driven product understanding with database-driven customer understanding has given rise to a radically different—and, some would say, radically improved—retail experience. Without a doubt, kiosks popping up in brick-and-mortar stores (e.g., Borders and REI) mark the return of such customer and product intelligence to physical retailing.

The cognitive prowess of other successful online ventures, especially the exchange businesses, is similar. How else might you characterize eBay's auction business, Match.com's personals business, or Monster.com's job placement business? Each company is primarily a business based on what databases do best. The databases behind eBay keep track of over 330 million listings a year and peer-to-peer ratings of over 100 million buyers and sellers. Match.com's databases manage the profiles of millions of men and women seeking men and women. Monster.com's databases keep tabs on job recruiters with positions to fill and individuals (and their résumés) looking for jobs. The genius of these businesses lies in database overlays, sources of customer value that could not generate sufficiently meaningful outcomes or create truly efficient markets at scale without the Internet. In more complex exchange businesses, such online brokerage sites as Charles Schwab & Co. and E*Trade, activities extend beyond self-contained databases. Why? To provide their functionality, brokerage sites must link to preexisting legacy organizations of people and machines executing transactions on physical trading-room floors, like the New York Stock Exchange, and electronic trading systems, like NASDAQ. The same is true of the mainstream travel sites, such as Expedia and Travelocity, or the reverse-market travel sites, such as Priceline and Hotwire. Regardless, the Web made these businesses viable by linking vast numbers of users with industrial-strength machine intelligence, using databases to enable the vast majority of cognitive processes.

This kind of database-driven cognitive superiority of machines has enabled Google, the Web's premier search engine, to threaten Yahoo!, the most trafficked portal on the Web. Yahoo! has relied historically on search algorithms built by people; Google relies on analysis of search patterns compiled by machines. Processing archival data on Web traffic, Google has built a smarter database, predicated on tracking and analyzing usage of the Web by the very users it proposed to serve. In doing so, it has largely cornered the market on search. Primarily by selling links from search results to online merchants and advertisers, Google generated revenue of $902 million in 2003, with 2004 revenue estimated at over $1.8 billion. Gross margins for both years exceed 85 percent. Google's market capitalization upon its IPO surpassed $23 billion.[20]

Cognitive capabilities of machines frequently provide the basis for satisfying interactions with technology-mediated interfaces. Think of the delight that Google users have experienced when they get a glimpse of the site's seemingly endless array of raw intelligence: Type in a FedEx or UPS tracking number, click the Search icon, and Google takes you to tracking information for your package; type in a ticker symbol, and Google generates the relevant stock market quote; or type in a vehicle identification number, and Google provides make, model, and year for the car. Those are just a few of the site's specialized applications of machine intelligence. You can also type in a numerical equation, a unit of measurement, or a name, and Google becomes a calculator, a converter of units of measurement, or a phone directory.[21] So, the next time you find yourself dealing in vain with the voice prompts at your Visa or Master Card service center and decide that you hate technology, please note: the problem is not the automation, but the inadequate cognitive capabilities of the machines your credit card issuer has deployed. Many such systems are under qualified, in cognitive terms, for the jobs they hold. Interactions get worse, not better, when voice-recognition software is added to the interface of many VRUs. As interface expert Jef Raskin has observed, the imposition of natural language interfaces does little to change the nature of interactions.[22] An inadequate or infuriating machine will perform inadequately and infuriatingly whether you interact by voice or keypad. The most satisfying machine interfaces to use are often not the most attractive but the most intelligent and intuitive. For example, if the VRU's virtual service representative at your local Poland Spring water delivery line—who introduces himself by saying, "You can call me Tom"—were as smart as Google, he would not garble everything you tell him and ultimately refer you to a call queue where you wait for hours to reach a live operator who still knows nothing about you, even though you entered your account number several times. If Poland Spring's Tom were as smart as Amazon, he would learn your preferences and needs, and become more compelling and intuitive every time you called back. Without the cognitive dimension, however, Poland Spring's Tom is just a dumb machine—worse, a dumb machine whose interface implies human intelligence—and dumb machines serve people poorly, even (and especially) though they sound friendly, behave respectfully, and speak customers' language perfectly.

The Emotional

Before you decide that this next element of machine interfaces takes the argument one step too far, we should acknowledge an obvious fact: So far as we know, machines have no emotions. They may malfunction, they may be sluggish, they may seem punitive. But outward appearances have nothing to do with actual realities. When we talk about machines having "emotional attributes," it's all in *our* eyes. We are projecting these attributes upon them, when they appear to experience emotions or when they elicit emotional responses from us. The same could be said of cognitive attributes ascribed to machines in this chapter. For example, when we talk about the cognitive capabilities of machines, we are not actually suggesting that machines think with a sense of identity and consciousness as humans do; rather, we argue that they draw on processing power such that they appear to think in human or superhuman ways. One day, we may build machines that do have emotions—or, as Moravec argues, that experience some primitive forms of consciousness. This may become a conundrum intrinsically related to the challenge of developing machines with humanlike intelligence. As Marvin Minsky, in *The Society of Mind*, observes, "The question is not whether intelligent machines can have any emotions, but whether machines can be intelligent without emotions."[23] But whether machines have emotions matters little to our discussion of customer interactions and relationships. After all, perceptions are most of reality. Who cares whether machines actually think and feel when they interact with human users, as long as they appear to think and feel—and that is part of what evokes our emotions.

We previously noted Honda's obsession with how its cars feel to users. What if a company competed on how products made customers feel emotionally? That's what companies such as Sony and Toyota aspire to do. In April 2003, Sony's CEO, Nobuyuki Idei, announced a new line of consumer electronics and computing products that, like Walkman or PlayStation, would come to market under a new sub-brand. The line was called Qualia; the word referred to the kinds of feelings Sony intended for its consumers to experience when they interacted with these products. Idei explained that the concept of Qualia was what people felt when they entered a room of family and friends who welcomed them with open arms. Idei proclaimed that consumers "will have delight in just holding these products."[24] At about the same time, Toyota released its first major redesign since 1996 of its best-selling car, the Camry. Press conferences for the new Camry focused on how the car felt to own and drive; Toyota distinguished this one from earlier Camrys by its emotional, not technological, appeal. Toyota engineers explained, "It's not enough to compete on performance anymore. The new competition in cars is on emotion." Not coincidentally, the Camry redesign garnered Toyota's largest media campaign in the United States—a $160 million cross-platform campaign centered on a musical theme and tagline "Get the feeling" (which replaced "Toyota. Everyday.").[25]

In parallel with the Camry announcement, news stories began to appear about another device that appealed to users' emotions—so much so that some began to anthropomorphize it by attributing to it humanlike intelligence, judgment, and personality. That device was TiVo. Perhaps the most personable of the so-called personal video recorders (PVRs), TiVo is a digital device, roughly the size of a VCR, that not only locates TV programs you care to watch and records them digitally on a hard drive, but also can observe what you watch and retrieve other programming that it believes will appeal to you. Some TiVo owners who use this feature experience strong emotional responses to their TiVos' conclusions about them. For example, a number of single male TiVo owners became unsettled when they realized that their TiVos had decided they (the owners) were gay, which they inferred from TiVo's programming recommendations based on their viewing habits. Many of these TiVo owners believed that they should set the record straight (as it were) by choosing compensatory programming that would "change TiVo's mind." By viewing cop dramas and Schwarzenegger movies—and no Lifetime Television or HGTV—they ensured TiVo's accurate reassessment of their orientation.[26]

TiVo encourages such attributions of free will to its PVR device by using a logo, which is a small stick-figure character with a TV-screen-shaped potbelly. TiVo is what computer scientists would call a service robot, and many TiVo owners think about their PVR as such. For instance, they speak affectionately about the device as if it were a loyal household pet that fetches TV programs instead of slippers. Of course, TiVo is not alone in anthropomorphizing its product or brand to engender an emotional response among customers. Many companies attempt to win over customers on an emotional level by associating their brands with human and animal personalities. There are the strong men (Mr.

Clean and the Jolly Green Giant), the appealing homemakers (Betty Crocker, Sara Lee, and Mrs. Folger), the reliable helpers (Mr. Goodwrench and the Maytag repairman), the appealing restaurateurs (the late Dave Thomas of Wendy's and KFC's Colonel Sanders), and the indefatigable service providers (Verizon's network tester and the Energizer Bunny). The emotional appeal of human or animal forms in media is time-tested throughout marketing practice. Adding personality associations to products can induce greater loyalty. When Volkswagen launched the New Beetle, it emphasized the personality of the car, implicitly carried over from *The Love Bug* days, when an old Beetle starred as the protagonist of the 1969 Disney film. When Nissan launched its Micra, its advertising urged customers to enter the car's special world by learning to speak Micra, the Micra's own secret language. This same marketing logic has made Japan's Hello Kitty franchise a brand sufficiently powerful to sell a wide range of merchandise around the world, raking in $500 million in revenue for Sanrio and billions more in license agreements.[27]

When marketers seeking to establish richer interactions with customers combine the emotional appeal of human and animal personalities and characters with machine interfaces, curious things happen, such as Tellme, the over-the-phone directory service that uses voice recognition and synthetic personality to furnish information to callers about movies, restaurants, stock quotes, and the weather. While Ananova, the virtual personality who delivers headlines on the Web, simply tells you the news, Tellme interacts with users. Tellme provides callers of 1-800-555-TELL with an automated interface that sounds like a friendly assistant; she takes verbal requests and not only retrieves relevant information but also offers to perform tasks on command, such as making restaurant reservations. Tellme also operates more business-oriented services—the front end of the automated system runs AT&T's toll-free directory assistance services—and the progeny of Tellme surrounds us.[28] One of us called to renew a magazine subscription, only to hear a VRU that said, "You've reached Susan, your automated renewal assistant. Are you calling to cancel your magazine? Please say 'yes' or 'no.'" By the end of that conversation, we *liked* Susan.[29]

In the 1970s, Joseph Weizenbaum sought to discredit such "artificial intelligence." He created the software program Eliza to demonstrate that even pure logic could generate "emotion," by engaging in Teletype dialogue with users and posing psychotherapeutic questions and offering ambiguous responses. Interestingly enough, Eliza's users knew that Eliza did not exist, but their interactions with her still comforted them.[30]

The Eliza phenomenon is consistent with more recent psychological research examining human responses to interactive technologies. In the early 1990s, two social scientists, Byron Reeves and Clifford Nass, began exploring the nature of interactions between humans and machines. Did people react to machines entirely differently from how they reacted to people—or did they react, in psychological terms, largely and predictably similarly? To generate findings, the PCs were set up with simple programs designed to express textbook human personality types while performing routine tasks or playing simple games. The hypothesis was that machine-mediated personality types, when matched with humans of certain personality types, would elicit classic responses in humans as already documented in the literature. The research showed that humans do not merely anthropomorphize machines; they anthropomorphize them in highly specific and very human ways, based on experiences acquired through previous social interaction. To demonstrate this point, Reeves and Nass identified two dozen of the most influential studies in the social science literature that dealt with people responding to each other and their natural environment; these studies examined how people of different personality types, ages, and backgrounds responded to other individuals or their environment given different emotional and social contexts, personality types, and environmental controls. Reeves and Nass restaged each study, but with a critical substitution. In each case, they replaced one side of the person-to-person dyad with a machine, for example, using a PC running simple software designed to play a game or perform a task while exhibiting certain personality types associated with specifically indicative behaviors. Test subjects would use their PC, for example, to play a Twenty Questions-like game. Each of the computers was programmed to offer praise, criticism, or no feedback at all while playing the game. As when interacting with other humans, respondents liked being praised by the computer and thought more highly of computers that praised them (even if the praise was unwarranted), while disliking computers that criticized them. Through a series of experiments like these, Reeves and Nass concluded that when humans engage with interactive machines—even generic PCs or television sets lacking physical attributes resembling humans or animals and without pronounced cognitive attributes—they react emotionally and predictably so, based on the human responses reported in the original studies involving human-to-human interactions. The machines in these experiments were merely running software, free of emotional expressions or intent. But they managed nonetheless to render the affective responses that belong, in theory, uniquely to the province of human social interaction.[31] The phenomenon is what Reeves and Nass call "the media equation."

Here is a fascinating paradox: For the human test subjects, the interactions with the PCs consisted of trivial undertakings that were largely devoid of meaning, yet the human emotions those interactions aroused were real and sometimes powerful. It's analogous to how we feel when a store clerk treats us rudely or proffers an unsolicited compliment; though the interaction is insignificant, and the clerk is someone we'll likely never see again, the emotions we experience are real—and sometimes powerful. Interfaces that engage on emotional dimensions have a unique purchase on customers and their attention. That is why service quality disproportionately influences purchase decisions, dominating other more rational elements that marketers call "buyer purchase criteria," and why world-class service providers toil to perfect the attitudes and behaviors of frontline service workers in customer interactions. Since we already have emotional relationships with products in our lives, many of which are becoming richer thanks to embedded smart technologies, we are psychologically prepared to accept machine-mediated service interfaces that appeal to us on emotional levels. As media consumption shows, consumers around the world are migrating from one-way experiences, such as TV viewing, to interactive experiences, like being online.[32] All of this signals a readiness among consumers for greater degrees of engagement and literacy with machines—and an acceptance of interactions and relationships with technology-based service interfaces not only on cognitive but emotional dimensions as well.

The Synaptic

While the connectivity that links machines and people is something that we've talked about extensively, there is another synaptic attribute of machines that's real today and growing by leaps and bounds. It's pervasive computing, or machine-to-machine communication. From one perspective, there is nothing new about this concept. Smart machines have been talking to other machines for decades, from time-sharing systems with terminals talking to mainframes, to broadcast-fax machines using phone lines to contact recipient fax machines around the world. Indeed, data traffic—namely, machines talking to machines—on the world's telecommunications networks overtook voice traffic nearly two decades ago. From another perspective, however, there is something radically new here. It's the capacity for machines of all kinds, not just computers and peripheral devices, to do more than merely communicate with one another; it's the capacity to build complex systems of devices that perform in ways that are explicitly predicated on the synaptic connections of machines to one another. Many of these innovations result in remote machine labor that supplants proximal human labor partially or even entirely. For example, remote meter reading uses wireless networks to enable meters to report to utility companies on power usage without human intervention. The alternative, which persists in many communities today, is for employees to work as meter readers for the power companies, driving from house to house and knocking on doors to gain entry to collect the information manually. While this application may be a leading example of network-based machine-to-machine communication, and the labor substitution it enables, there are tens of millions of other devices of this kind talking to one another in similar ways.

One research organization predicts that by 2005, machine communication over wireless networks will surpass voice communication. To illustrate why, consider an Atlanta-based supplier of swimming pool chemicals. The company, BioLab, has developed a system that keeps watch over the water quality of its clients' pools. Its remote sensors report data to a central processing facility at the company's network operations center in Chicago using the cellular network, where the data is analyzed and stored. If there is a chemical imbalance in the pool water, the network calls or pages a third-party maintenance person in the area who has been retained by the customer. If that person takes no action, the system contacts the next person up the chain of command, the maintenance person's boss, until action is taken. Initial estimates suggest that a $2,000 investment in the system for a single large swimming pool can result in savings of $3,000 a year in chemicals to maintain pool operations, based on a higher quality of monitoring.[33] Though the application could work without wireline connections, wireless technology results in a more flexible system that operates at lower cost. In this case, one by-product of machines talking to machines is, incidentally, the use of machines to provide managerial oversight of people—the maintenance workers.

Integrating the Attributes

When machines can excel with respect to several of these attributes—the physical, the cognitive, the affective, and the synaptic—simultaneously, the media equation becomes more profound in its implications. The experiments of Reeves and Nass tested only one dimension of human-to-machine interaction—namely, response to visual stimuli. The researchers did not attempt to differentiate the machines in terms of richer attributes or even rich visual media—the machines were programmed with basic software to express symbolic personality types. It's no wonder that products such as AIBO, TiVo, and Roomba have caused such curious responses of attachment among their owners. Each exemplifies an experiment with a device that delivers on several attributes simultaneously; AIBO is arguably the most advanced (with significant physical, cognitive, and emotional development) and Roomba the least (with some physical and limited cognitive development). TiVo uses colorful and humerous graphics (affect) and network support (synaptic connections) to build richness in cognitive interactions. Each of these machine-dominant interfaces forms bonds with owners the way people or household animals do. From a psychological perspective, the impacts of such products on customers transcend anything that's come before in consumer packaged goods or consumer durables. This is, indeed, a new frontier of human experience in both the commercial and social spheres.

To get a sense for the power of machine interface attributes working in concert, consider the case of a start-up company from the 1990s called Wildfire Communications. Our work with Wildfire began nearly a decade ago, when it was just another high-tech start-up outside of Boston. Its founder, Bill Warner, wanted to start a company that would "make the most popular software in the world." Observing that the telephone needed a new interface (touch-tone was arguably the only major interface innovation since the telephone's invention in the late nineteenth century), Warner aspired to change how people interacted with telephones and telephone networks. At Wildfire, he and a band of engineers developed a system that combined the functionality of a receptionist, a Rolodex, a switchboard operator, an administrative assistant, a conference coordinator, and a voice-mail system; the user would manage all this functionality through interactions with the Wildfire assistant, a personality with a human voice that talked to you through the phone and responded, Tellme-style, to voice commands.

In the early days of the company's history, we suggested to colleagues who were researching service businesses at Harvard

Business School that they try Wildfire for a year. Not everyone found Wildfire, as the virtual assistant was called, compelling: Two colleagues ultimately took little interest in "her" and remained heavy users of the school's voice-mail system, one used Wildfire lightly, and a couple of us became heavy users. If you really got to know Wildfire, she turned out to be an extremely appealing presence in your life. She screened your calls, figured out whose calls you generally wanted to take, return later, or ignore; she let you know who else was in the system when you came in (a feature called Virtual Hallway), and she kept track of a mind-boggling array of contact numbers for everyone you knew. Best of all, if you were driving down the highway cradling a cell phone on your shoulder, she could take commands and get things done without your ever having to take your eyes off the road. (This is what made Wildfire an attractive acquisition for the European mobile-phone operator, Orange, a unit of France Telecom, which acquired the company in July 2000.[34])

When our little experiment ended, we observed wildly varying emotional reactions among colleagues. The nonusers hardly noticed Wildfire's absence; light users missed her a little bit; and heavy users suffered the pain of loss. We discovered that Wildfire had impressive cognitive and synaptic capabilities—like a talented receptionist, she could track down and connect people to one another—and emotional functionality. Normally, when you called her, "Wildfire!" she would say, "Here I am!" When you called after midnight, she would first sigh and then, in a less energetic voice, say, "Here I am." If you asked her, "Do me a favor," she would answer, in a slightly inappropriate tone, "What *kind* of favor?" If you replied, "What does a cow say?" she would respond, "Mooo!" Our favorite Wildfire experience, however, occurred on a hectic Friday afternoon in a crowded airport. One of us, stranded for the evening and seeking a hotel room, called Wildfire to check messages and return calls:

"Wildfire!"

"Here I am!"

"Do me a favor."

"What *kind* of favor?"

"I'm depressed!"

"You think *you're* depressed—I live in a box!"

Her reply made our day. When you have a witty "friend" like that—who works tirelessly on your behalf twenty-four hours a day—how can you not grieve when she leaves? Having Wildfire unplugged led to a period of emotional loss that was real, even though she was assuredly not. When you live through such experiences, you realize that interactive technology, in its cognitive and affective forms, is not a tool or a machine as previous generations have understood and defined such things. Machines that create meaning and value in interactions and relationship with humans, especially in consumer mass-markets, are altogether new. They are also unlike any form of machine-based automation that's come before.

WHAT WILL THE FUTURE BRING?

How soon will machines or systems combining behaviors, thoughts, and feelings exhibit more developed human traits is open to question. As noted, Ray Kurzweil has argued that we are at the brink of "the age of spiritual machines." His prediction is based on the idea that thoughts and feelings are not so much ineffably or spiritually unique traits of human life but the direct result of human-like levels of information processing power. Such processing power in machines is rapidly closing the gap with the capacities of the human brain. Kurzweil points out, "The human brain has 100 trillion connections, each computing at 200 calculations per second, which means that humans can make 20 million billion calculations per second. In 1997, a $2,000 computer with parallel processing could perform 2 billion connections per second." He argues that with processing power accelerating according to Moore's Law, by 2025 this technology evolution will result in a $1,000 PC with the capacity of the human brain.[35] At that point, machines will develop consciousness—they will become spiritual—as a function of their elevated intelligence. Others, like roboticist Rodney Brooks, believe that humans will not be overtaken by machines but rather, that we will merge with them by adopting chip and device implants that can interact directly with our brains and nervous systems. In this way, Brooks believes, humans will always remain one step ahead of technologies that might otherwise overtake us in depth of thought and feeling.[36] Hans Moravec, who laid out the four generations of robotic evolution we cited earlier, has long argued that silicon substrates represent the next phase of human evolution. For many years, he has contended that the next great stage in human evolution will entail a transfer of human life and consciousness from its basis in carbon-chain strands of DNA to silicon-based chip intelligence.[37]

But regardless of the scenario for now, we must ask: What are the business implications of machines that can increasingly interact, relate personally to, and satisfy customers on behalf of companies? In this regard, the role of emotion deserves particular note. The inability of machines to appeal emotionally to customers has for many decades limited their roles in business to back-office and production tasks. Just as interpersonal or professional skills distinguish white-collar from blue-collar roles among human workers, machines with the ability to relate to human beings on affective dimensions are the ones graduating to front-office roles. Until now, few if any machines had the complex array of interpersonal skills to perform such jobs. Emotion was the missing piece. Computer scientist Rosalind Picard writes, "Emotion has a critical role in cognition and in human-computer interaction. Computers do not need affective abilities for the fanciful goal of becoming humanoids [but] for a meeker and more practical goal: to function with intelligence and sensitivity toward humans."[38] Computer scientists and programmers have ignored, in her view, this fundamental aspect of human intelligence in their development of machine analogues and interfaces. Emotions represent an unacknowledged challenge among computer scientists attempting to render or augment human attributes in machine code. Congruent with the findings of Reeves and Nash, Picard observes, "Human-computer interaction has been found to be largely natural and social; people behave with computers much like they behave with people."[39] She concludes, "There is a time to arouse the passions of others, and a time to diminish them; a time to decide with one's head or with one's heart, and a time to decide with both. In every time, we need a balance, and this balance is missing in computing."[40]

Of course, there is a dark side to the realization of Picard's challenge—or ours. In *The End of Work*, Jeremy Rifkin predicts that we'll reach a point in 2050 when machines are capable of performing more than 95 percent of the meaningful work required by the world's economies. In his manifesto, he warns that we humans, as a species, will endure enormous social dislocation because the world we inhabit will no longer need us except as consumers of industrial and service outputs.[41] Rifkin comments, "The great issue at hand is how to redefine the role of the human being in a world where less human physical and mental labor will be required in the commercial arena. We have yet to create a new social vision and a new social contract powerful enough to match the potential of the new technologies being introduced into our lives. The extent to which we are able to do so will largely determine whether we experience a new renaissance or a period of great social upheaval in the coming century."[42] Rifkin's Jeremiad resonates with another concerned commentator whose focus is primarily on the threat of robots to human life and well-being. Marshall Brain, a Web site publisher and nonfiction author, believes that 5 million jobs in the retail sector will be lost by 2015 as a result of automation.[43] His thesis is simple and plain: "The rise of the robotic nation will not create new jobs for people—it will create jobs for robots." Brain proposes social policies that would call for governments of the world's major industrial economies to furnish their citizens with $25,000 a year in state-funded stipends, with the goal of ensuring that idleness resulting from robots having assumed the majority of jobs does not lead to poverty or social upheaval among the masses. (Brain predicts that by 2030, robots will have eliminated employment for some 50 million Americans, mostly among those 60 percent of the work force who are employed in jobs paying $14 an hour or less today.) Through such income redistribution, Brain believes economies like the United States' can sustain the consumer spending power essential to continued economic growth and development; the alternative he foresees is a chronic global recession.[44]

Even as these prophets of doom deliver their messages, we have in the past few years begun to experience our dependence on technology in ways that are markedly discontinuous with the past. The apocalyptic vision of computers or robots taking over is only slightly more alarming than what happens to society or the economy when technology fails or proves vulnerable. Such potential for catastrophic failure has been highlighted by the shutdowns of electric power grids in the United States and Canada in 2003; hacker attacks targeting online sites or security flaws in the Windows operating system; information security threats to governments and corporations; and widespread flurries of viruses and spam launched over the Internet. Rifkin and Brain focus on the threat to jobs and, ultimately, the meaning of work and life. We might also focus on the perils of becoming too dependent on technology, even if it only plays enabling roles in human life and work. Such polemics notwithstanding, however, the question remains, What should business do? In the next few chapters, we'll explore potential answers. In our view, the outcome will be far more complex than a substitution of humans by machines, of robots for people. There is too much evidence already—including the need for human safety nets to address the recurring failures of large technological systems—that hybrid forms involving the collaboration of people and machines will characterize the future of work and commerce. The machines will not take over, unless we want them to; and our success will depend on making certain that technology, which is neutral, bends to our vision of what reality should be.

CONCLUSION

The phenomenon we've been examining in this chapter defies characterization as simple automation, and its impacts will hardly be limited to efficiency gains. Given the scarcity of appropriate talent in the face of businesses' rising demand for front-office or relationship management positions, machine-dominant service interfaces must inevitably play a central role in the future of business. When corporations must deliver interactions and relationships consistent with the personality of their brands and with favorable economics, machines will substitute or complement human labor, depending on the nature of the task. The deployment of machine-dominant interfaces implies that:

- *Machines excel at certain kinds of tasks relative to human capability and not at others.* Machines excel at providing immediate "always on" access to vast quantities of information, accuracy and precision in rote or repetitive tasks, reliability and indefatigability, and the mediation of personalized interactions drawing on database resources. People are best equipped to respond to the unexpected with flexibility and creativity, to express warmth, to exercise judgment, and to deal effectively with complex interpersonal situations.

- *Machines may present themselves increasingly as having personality and emotions, but their functionality must back up the human impression.* When machines mimic human qualities, they can prove endearing but they also raise expectations regarding the complexity of what they can do. When machines seem human but perform tasks poorly, they increase frustration as a function of their having elevated customers' expectations.

- *It's clear how machines create efficiency, but they must also deliver effectiveness.* Through automation, machines in front-office roles can drive down operating costs. But frontline positions also create or destroy upside for businesses, based on the quality of interactions or relationships they mediate with customers. Managers must not settle for mere automation by front-office machines; the machines must interact in ways that build customer perceptions of value and drive increases in customer satisfaction and loyalty. Otherwise, the machines have helped companies win only half the battle and that half, in most industries, will qualify as table stakes.

- *Machine-dominant interfaces are not, and will not soon become, fail-safe.* Machines excel at rules, not exceptions. In addition, systems fail. Since customer interactions are inevitably subject to the vagaries and exigencies of human life, not every customer contact will fit with automated relationship management. And no automated interaction will work flawlessly. As a result, managers must establish in advance the default modes for any machine-dominant interface. These may involve referrals to other machines, but they will likely require links to skilled human beings.

- *The appeal of front-office machines is often more obvious to managers than to customers.* Because there are tangible cost advantages to deploying machines in the front office, managers run the risk of embracing front-office automation well ahead of their customers. It's difficult to force customers to accept machine-dominant interfaces unless customers are ready to accept them. Machines must make themselves welcome in a social context before front-office reengineering can generate real returns.

- *Technology is a moving target.* The evolution of machine intelligence, including the acquisition of emotions and judgment, will accentuate the trend toward front-office reengineering. It will also alter many considerations regarding what machines do best and what people do best. At the same time, the diffusion of front-office machines will increase customers' readiness to accept new kinds of interfaces, as our collective mass-market experience of the Web has already done. Even as advances in technology change some of the dynamics regarding implementation, the underlying themes of this chapter— and especially the four interface attributes—will become important areas for managers to monitor over time.

6

Putting the Amalgam
of People and Machines
to Work

A s we have seen, people-dominant interfaces have long pre-
vailed in traditional frontline services. But technological evolution and customer readiness among other factors are now
enabling deployment of machine-dominant interfaces on the front
lines of many businesses. Each approach has its strengths and
weaknesses. People excel at conveying empathy and handling exceptions but are challenging to manage and costly to deploy and
train, especially in large-scale service operations. Machines excel
at processing information and performing rote or repetitive tasks
but can depersonalize or homogenize interactions. In effect, the
front office needs machines to compensate for people's shortcomings and people to compensate for machines' shortcomings.

That's why we believe that tomorrow's mainstream service
interface will be hybrid—one that creatively amalgamates the
strengths of people and machines (see figure 6-1). In structuring
such innovative interfaces, businesses will deploy two variants of

FIGURE 6-1

What Humans and Machines Do Well

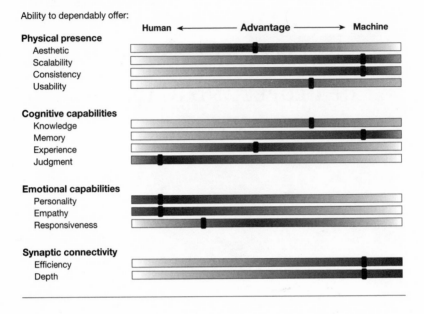

Ability to dependably offer:

Human ←———— Advantage ————→ Machine

Physical presence
Aesthetic
Scalability
Consistency
Usability

Cognitive capabilities
Knowledge
Memory
Experience
Judgment

Emotional capabilities
Personality
Empathy
Responsiveness

Synaptic connectivity
Efficiency
Depth

the hybrid interface archetype: one where people operate in the foreground and are supported by machines, and one where machines operate in the foreground and are supported by people. We call these people-led and machine-led hybrid interfaces, respectively. Such hybrid models optimize the trade-offs between efficiency and effectiveness in customer interaction and relationship management. Of course, many hybrid interfaces will involve people in the foreground, enabled by machines in the background, enabled by still more people and machines. Even though such multilayer or multilevel interfaces may become commonplace, this chapter deals with the fundamental building blocks of complex interfaces—the hybrid forms.

The reality of hybrid interfaces can seem by turns wildly innovative and yet comfortingly familiar. For example, many years ago, the MIT researcher, Steve Mann, tried living his life in constant communication with the world's knowledge networks. Beginning in 1982, before the laptop and Wi-Fi, Mann assembled a portable PC to wear on his back and a "heads-up" display over one eye, with a chorded keyboard in his pocket that enabled him to enter data and commands with one hand, and a wireless network that connected his equipment continuously to the Internet. If you began discussing an obscure subject with him, he might have little to say initially; but, after accessing data online, he could speak knowledgeably about your topic.[1] Mann's latest wearable computing system, WearComp7, consists of a seemingly ordinary pair of sunglasses functioning as a heads-up display, connected to tiny electronic components hidden in his clothing that supply computing power, memory, and wireless connectivity.[2]

What started as a wild experiment in an MIT laboratory has become reasonably mainstream in business today: the handheld tracking and scanning devices of FedEx delivery personnel, the small computers appended to the belts of Staples or Target clerks for accessing product information, and the wireless phone headsets that clerks at retailers like Old Navy and Banana Republic wear. The upscale London-based restaurant chain, Wagamama, specializing in New Age Asian noodles, utilizes the hybrid approach. In the dining area, servers take customers' orders on handheld PDAs with wireless connections to the kitchen. If customers order everything at once—food and drinks—their beverages may arrive before the server taking their order even leaves the table.

A recent advertising campaign for BlackBerry—the handheld device that acts like a PDA with e-mail—juxtaposes the human-dominant and the hybrid interfaces. In the first scene, the service person in the maintenance hangar tells the owner of a private jet that the part he needs *may* be in the distributor's inventory. In an alternate scenario, the service rep uses a BlackBerry to confirm that the part is available and will arrive the next morning.[3] The ad captures the powerful union of the strengths of humans (in this case, face-to-face empathy and reassurance) with those of machines (access to real-time data in far-flung complex systems).

Such technology-enabled interfaces can transform average workers into heroes and stars.

The reverse of the aforementioned hybrids is machines enabled by people in equally wild, yet familiar incarnations. Remember that odd creation of the Depression Era in New York City—the Horn & Hardart Automat? It was an ostensibly "modern" service, devoid of frontline service workers except for the lone woman who sold tokens in a booth inside the restaurant's entrance, with walls of windowed metal boxes displaying food and beverages that customers could open by inserting special tokens in the slots. Behind the boxes, you could glimpse a bustling industrial kitchen, full of white outfits replenishing the windows emptied by hungry patrons. Like vending machines today, the interface consisted of machinery but was enabled by people.[4] Contrast that with the now familiar experience of dialing directory assistance. A recorded voice answers you by asking for the city and listing. Once you reply, a live operator actually handles the call, unless you happen to connect with a fully automated system such as Tellme. You may or may not hear the operator's voice as the operator identifies the number and authorizes another recording to recite the number and thank you for calling. For continuity between spoken machine prompts and live conversation, operators often record their own voices to intone greetings and commands, whereas voice-synthesizing chips generate the phone numbers. By unbundling the call structure—where standard components are automated and custom components are delivered "live"—operators can handle more calls per hour and speak fewer words. Idle chitchat decreases, and productivity increases.

Perhaps the starkest contrast of the two types of hybrid interfaces comes from the U.S. Department of Defense in its deployment of two attack planes, the Stealth Bomber and the Predator Missile. The Stealth Bomber is so complex and so unwieldy that no human being or crew of humans can keep it aloft, let alone pursue a target, while attending to its many systems. The Stealth is like any commercial aircraft in that on-board computers control many of the variables that enable flight. But, where commercial planes can be flown without machine assistance, the Stealth cannot. In contrast, the Predator is actually an unmanned aircraft that can fly repeatedly into hot war zones to attack military targets without risking a pilot's life. The Predator's pilot maneuvers the plane from a command module that picks up in-flight sights and sounds from sensors built into the aircraft.[5]

In aviation, these two approaches to hybrid interfaces—pilots supported by machines (Stealth), and machines supported by pilots (Predator)—rely on radically different architectures. This chapter focuses on the business applications and implications of such hybrid interfaces. We will look first at people-led interfaces, then at machine-led interfaces, and finally at the best practices of a U.K. bank that successfully used a variety of hybrid interfaces to reinvent retail banking.

PEOPLE ENABLED BY MACHINES

What constitutes a machine? To broaden our perspective, consider Wal-Mart, now the largest retail chain in the world, selling products ranging from consumer packaged goods to packaged media and an employer of more than 1.5 million people.[6] Recent studies reveal that 25 percent of U.S. productivity gains in the late 1990s came from increases in efficiency at Wal-Mart alone.[7] That resulted from Wal-Mart's retail automation, along with its purchasing power and channel muscle. Its stores may look low-tech, but its operations are high-tech: innovations such as efficient consumer response, category management, and cross-docking yield enormous distribution efficiencies. Wal-Mart puts a human face on retail distribution through its many store associates and greeters, typically senior citizens who embody such small-town qualities as kindness, warmth, and hospitality. Its greeters welcome you when you enter and thank you when you exit, so that you have positive first and last impressions (which weigh most in determining customer perceptions of service) within a highly automated environment.[8] In essence, Wal-Mart is a machine with a human face.

But Wal-Mart's retail personnel need not understand the machine's inner workings. Retail clerks hired for their human qualities can focus on behaving warmly, considerately, and respectfully. Other personnel can deal with mechanics behind the scenes. This division of labor has backed many retail formats. For example, teenagers—not adults—operate the typical suburban shopping mall in the United States today. Rarely will you see an adult employee on the retail floor or in the back office of an Abercrombie & Fitch or a Best Buy. Unlike mom-and-pop retailers, these chains have control systems like Wal-Mart's: regional offices or headquarters, not staff in the stores, determine and centrally control merchandising, inventory management, product displays, pricing, discounting, and other aspects of store operations. Such centralization allows national chains to employ teenage clerks for their friendliness, pulchritude, and style, not for their retail systems experience, product knowledge, or merchandising skills.[9] Through machine systems, chain retailers can use energetic or attractive lower-cost labor to deliver distinct, live, branded interactions. Of course, the system grows more complicated when brands attempt to integrate the qualities of people and machines more tightly.

Machines Speaking Through People: Ritz-Carlton and Fairfield Inn

The efficiencies of mass-market retailers appear prosaic, if nonetheless impressive, beside those of the people-led hybrid interfaces when people and machines work in tighter integration. Ritz-Carlton, with fifty-seven hotel properties and resort locations worldwide, has experimented for several years with this configuration. While Four Seasons caters to elite society, Ritz-Carlton focuses on affluent business travelers. Its scale prevents Ritz-Carlton from relying solely on human resources to deliver personalized guest experiences; it uses technology to differentiate how its frontline staff serves guests. Facing commoditization at the high end of the hospitality market in the mid 1990s, Ritz-Carlton created a systemwide advantage through machines, specifically database technology. The chain invested in two systems for its properties—a national cross-property database of guest records called COVIA (named after the IT unit at United Airlines that originally developed the system) and a local property-specific customer database called Encore. The goal was to track the preferences of the chain's most frequent guests through observation and database technology to anticipate their future needs.

Consistent with the Ritz-Carlton brand, technology enables the front office but is largely invisible to guests. Upon phasing in the systems, management instructed each frontline service employee to collect information on the chain's most loyal customers. It used this data to populate the local Encore systems and then headquarters aggregated the multiple streams of Encore data into COVIA. To feed the system, every Ritz-Carlton employee must now carry a Guest Preference Pad, a small tablet of forms for recording guest observations, such as a guest's favorite type of pillow.

Ostensibly, little is new here. Airlines have long collected customer data in frequent flier programs. But only in the past few years have airlines used this information to generate individualized direct-marketing offers and to deliver targeted services to customers in real time. For example, when you reach elite levels of major airlines' frequent flier programs, you can access a range of privileges triggered by status markers in the airlines' databases, such as more liberal upgrade policies, priority on overbooked flights, and more lenient application of airline rules. More recently and less consistently, airlines have attempted to modify their service personnel's attitudes and behaviors according to a passenger's status in the system.[10] Of course, airlines process millions of passengers annually and deploy tens of thousands of frontline workers daily, whereas systems-based information in hotel environments can more effectively shape the behavior of frontline service personnel toward individual guests. Every morning at each Ritz-Carlton property, the staff is briefed on individual guests in the hotel whose preferences are in the system. Conceptually, they strive to accomplish three goals with each guest—observation-based personalization of service, anticipation-based customization of service, and service recovery, an important, though often overlooked, aspect of managing customer interactions. Making amends has enough value in economic terms to Ritz-Carlton that every worker on its payroll has a $2,000 budget to help a guest recover from a service problem right when something goes wrong.

The $2,000 per employee may sound dramatic, but the logic underlying this allocation of funds is impeccable. Loyalty research indicates that, all things being equal, the average rate of customer intent to repurchase is 78 percent.

- If a customer fails to complain, then the company misses the opportunity to address the problem; the customer is never made whole and will probably never return.

- If a customer complains to someone on staff and the company does nothing, then intent to repurchase actually rises—apparently, just getting a complaint off your chest can increase the chance of your returning.

- If a customer complains and the company responds slowly but effectively, then the likelihood of repurchase can exceed 78 percent.

- If a customer complains and the company acts quickly and effectively, then the customer will likely become more loyal to the company than if nothing had ever gone wrong.

Hence, some have called this notion "the profitable art of service recovery."[11] The logic of service recovery is embedded in Ritz-Carlton's service delivery systems, so much so that its staff members never refer to a guest's having a problem; rather, they will say, "Mr. Smith had *an opportunity*." Mr. Smith probably experienced something unfortunate and complained grievously. Ritz-Carlton's resulting business opportunity calls for the chain to respond so decisively that Mr. Smith will become more loyal than before. But a company cannot build such loyalty without human workers authorized to address negative situations in real time on the front lines and systems to track incidents and service recovery in the back office.

Data-driven systems have effectively focused Ritz-Carlton on the quality of customer interactions.[12] Requiring employees to note guest preferences provided a powerful mandate to change their attitudes and behavior. Once the chain populated its data-bases, frontline service personnel could interact quite differently with guests. For example, the system could cue a front desk clerk to upgrade Mr. Smith to a suite, due to a recent "opportunity" at another Ritz-Carlton property. Enabled by the machines, she can right a wrong and thrill a total stranger, which is a satisfying experience *for her*. Not bad for a day's work.[13]

Of course, optimizing the outcome depends on fulfilling related requirements: Personnel must collect information subtly to respect the guest's privacy and handle customer information so deftly that Mr. Smith does not conclude that he's living in Eastern Europe before the Wall came down. Also, the Encore and COVIA systems need sufficient security and IT protocols to prevent abuse of guest information by hackers and misguided employees. Finally, data handling—from input to management to inferences—must operate with utmost integrity to prohibit erroneous recommendations or outright violations of privacy.

Not surprisingly, Ritz-Carlton's parent company, Marriott, has experimented with multiple hybrid interfaces. At its economy limited-service chain, Fairfield Inn, it has relied on a management system built around Scorecard, a device that measures the chain's performance, customer by customer and property by property, against specific drivers of customer-perceived value, corresponding to specific service standards. The typical one hundred fifty-room Fairfield Inn—conceptualized in the mid-1990s by reverse engineering the needs of traveling mid-level businesspeople or "road warriors" who care deeply about service standards along several critical dimensions—lacks a pool, a restaurant, and kid-friendly facilities. According to initial research, road warriors' loyalty rested disproportionately on six factors, including overall cleanliness of rooms, friendliness at check-in and -out, and the speed or efficiency with which transactions were handled. Scorecard itself, an inexpensive PC located at the front desk of every Fairfield Inn, gathered guest feedback while clerks processed their paperwork on checkout. The machine presented five questions in random rotation and asked customers to rate the property's performance on a 10-point scale. Roughly two-thirds of customers completed Scorecard, because it took no more than 30 seconds.

Scorecard data gave property managers a dynamic picture of how well they were meeting customers' needs. Since Scorecard linked results to individual employee performance, matching guests and their rooms with housekeepers who cleaned those rooms, the data factored considerably into reward and recognition programs. Every employee's compensation was calculated as a function of overall guest satisfaction with the property and individual performance based on the service dimensions that Scorecard tracked. The system created accountability among individuals on their performance with guests *and* with each other, because an individual's underperformance eroded overall bonus income for each property's team. So employees focused on two goals: Am I delivering on key dimensions that drive guest satisfaction? Are all teammates delivering appropriately on the collective goals? The system exposed slackers whom coworkers—not management—quickly rooted out. By attracting and retaining those best equipped to contribute in such an explicitly measured system, Fairfield began developing a higher-quality, motivated, and more team-oriented work force with higher morale and a more productive work place. The increased productivity of employees has translated into more vacation days and richer benefits, while preparing them to host celebratory events involving employees and customers. Fairfield Inn's employee turnover is minimal compared to the hospitality industry overall, and pay substantially exceeds that of the limited-service segment.[14]

The Fairfield system is not as sophisticated as Ritz-Carlton's: Fairfield uses machines as blunt quantitative instruments to shape employee attitudes and behaviors, whereas Ritz-Carlton uses them in subtle qualitative ways to enrich the interactions between the chain's frontline employees and its guests. In both cases, however, machines connect what customers want with what employees deliver and service innovations depend on hybrid service interfaces with people out front.

Marrying People and Machines: McDonald's

Field-based research cannot substitute for rolling up your sleeves and serving customers yourself. So we secured employment at a local McDonald's for a weekday evening in New England. In the realm of front-office innovations, the fast-food restaurant sector has done little more than computerize cash registers and install drive-thru windows. The result: More than one of every two fast-food orders in the United States are filled incorrectly.[15] How can a major industry survive with an error rate of over 50 percent? Because customers buy nearly half of all fast-food meals consumed in the United States through drive-thru windows, and only a devoted few will drive back to complain after discovering an error.[16]

We wanted to witness firsthand the effects of front-office re-engineering on fast-food operations. This particular McDonald's featured newly installed software designed to run the restaurant's frontline operations (order taking, money handling, and food assembly). Every one of the dozens of items on a McDonald's menu comes in countless variations—multiple portions (regular or super-sized fries? how many McNuggets?), flavors (which sauce? which topping?), and so on. Options have even proliferated in combination meals and Happy Meals, designed partly to simplify the order-entry process. Since labor is divided among those employees who take orders and handle money and those who prepare food and assemble orders, order takers must enter the details of each order into the system so that "expediters" can put the right food in the bag before it reaches the customer.

The new software system deployed a set of employee-facing interfaces—high-resolution, color touch screens driven by two on-site servers—to manage the complexity of the menu and to make relevant tasks simpler and more intuitive. Touch screens on registers, order-entry PCs, and money-handling stations featured icons with links to menu options and cues to ask customers follow-up questions about their orders. As a result, information flowed through the restaurant in symbolic and visual forms. Food preparers could see orders as they progressed through the queue. Assemblers could await their instructions from screens that showed orders from both the counter and the drive thru. Cash register operators received visual cues for money in and out as well as pictograms of possible combinations of change. Everyone wore wireless headsets to alert one another of problems and to hear interactions at the drive-thru window.

Fast food's menu and pricing complexity has grown to such an extent that humans cannot easily process the data deluge without erring. Despite the physical character of the business, most of what happens in a McDonald's involves information processing. Of the fifteen people working while we were there, only four actually prepared food. The rest worked in front-office functions— real-time information processing (orders, money, fulfillment) and customer interaction management (greeting customers, delivering food, thanking them for their business). Our crew of frontline workers did what people do best in such a setting (relate to other humans) while the machines did what machines do best (crunch numbers and move data).

By mid-evening, when traffic volumes were down and our skills were up, we were no longer automatons in a chaotic factory. We recognized that customers coming through the restaurant's two main service channels had very different needs and expectations. The drive-thru segment had no interest in relating to a human being (much as we tried greeting them exuberantly as they drove up). Rather, they wanted an efficient transaction that met certain functional criteria, such as speed, accuracy, and responsiveness. In contrast, walk-in customers wanted to relate to a real person behind the counter. One apparent regular, a middle-aged man who looked as if he had spent too many years at high-tech start-ups, came in late. He eyed us suspiciously, placed his order hesitantly, then declared, "Something here is not normal. . . . It looks as if this McDonald's was taken over by Genuity."

We learned a profound lesson here about personalization. Businesses often believe that personalization means making a particular interaction personal through the service provider (i.e., "Hi, I'm Bob. I'll be your waiter today!"). But it's not. Personalization should be about designing service interactions that address individual customers' needs and expectations. We call this the *personalization paradox*—the notion that a personalized interaction or relationship may not always be personal. For example, one customer filling a prescription at a pharmacy counter may wish to have a dialogue with the pharmacist instead of his doctor; another may desire an anonymous transaction that safeguards her privacy at the point of sale. The pharmacist provides "personalized" service by treating the first customer in a familiar chatty way and the second, with professional reserve and neutral efficiency. Each approach is personalized—but by design, one interaction is extremely personal and the other is coldly impersonal. When the pharmacist consciously delivers interactions of such contrasting character, she is personalizing service delivery. Our customers that evening at McDonald's self-segmented along similar lines. Transaction-seeking customers came to the drive thru. Relationship-seeking customers walked into the restaurant. Each segment valued completely different employee behaviors in the interaction.

Fast-food restaurants as well as hotels and drugstores often lack access to sophisticated individuals for frontline positions in their operations. Generally speaking, fast-food franchises recruit high school students who spend an average of four or five months in these jobs. (Turnover in fast food is 138 percent a year.[17]) Yet people in these restaurants—like the Ritz-Carlton hotels and the chain pharmacies—do information-intensive work: getting a customer's Happy Meal configured correctly is largely a data-management challenge to personalize an interaction (with tangible output) and meet individual customer needs appropriately. Information complexity in large-scale service businesses can become daunting. Managers in such settings must not only increase per capita output of frontline workers but also increase the quality of their interactions with customers. Machines enabling frontline workers can enrich customer interactions (e.g., with customer information at Ritz-Carlton), resulting in increased *leverage* for a company's human talent. Machines enabling frontline workers (e.g., with enterprise software at McDonald's) can also liberate workers' time and energy for customer interactions, resulting in increased *productivity*. Companies deploying such people-led hybrid interfaces can realize gains in efficiency and effectiveness in the delivery of services and management of customer relationships.

Machines Supported by People

The alternate hybrid form deploys machines in the foreground enabled by people in the background. Again, these hybrid interfaces create new sources of value in two ways. First, machines in the foreground can provide people with leverage by distributing their personalities in scalable ways, similar to how the media builds entertainment brands and celebrities. Second, frontline machines can increase people's productivity by channeling their work more efficiently, as call centers do by delivering customers to service providers rather than sending service providers to customers. For example, Web sites such as drkoop.com and drDrew.com broadcast personality on an interactive platform, delivering celebrities to potentially millions of customers and increasing the leverage of purported human talent. (Dr. C. Everett Koop, who is no longer affiliated with the site, was the Surgeon General of the U.S. in the 1980s; Dr. Drew Pinsky is the host of MTV's popular *Love-*

line, a talk show for teens featuring dating and relationship advice.) A service such as LivePerson—an outsourcing company that provides live online customer support on demand for third-party sites—directs customers to service providers for text-based or click-to-callback conversations, increasing the productivity of frontline workers who deliver the services.[18] In each case, machines enable substitution (the interface is not human but machine) and displacement (the human talent is not proximal but remote), which are the twin drivers of front-office reengineering.

The case studies and the integrated example in the following sections explore the effects of leverage and productivity of hybrid interface designs where machines are supported by people.

Leverage and Productivity Through Technology Interfaces

If you have listened to the weather report on a popular radio station in a major city, then you have probably experienced the leverage of the hybrid interface where a machine—your radio—sits in the foreground. In Boston, for example, most denizens will recognize the voice of former local radio personality Joe Zona, known over many decades for his weather reports. Zona had the newsreel-style voice of early radio days and the brisk cheer of a man on the go. A few years ago, we observed Zona at work: He recorded his local forecasts from a tiny sound studio in the basement of a Victorian house in Bedford, Massachusetts. The facility was operated by the nation's foremost source of weather-related information, Weather Services International (WSI), a little-known company that employs dozens of skilled meteorologists. WSI prepares the weather page for every daily edition of *USA Today*, the largest-circulation daily newspaper in the United States, and it furnishes crop reports and long-term weather predictions over squawk boxes linking its professionals to commodity trading pits at the Chicago Mercantile Exchange. In addition, WSI analyzes weather data to determine government and school closings across the United States during winter storms and maritime decisions on boat movements around the world. The company also generates thousands of weather reports for U.S. local and national media in every region of the country. Given the scope of WSI's business, we suspected that Zona's reach went far beyond Boston. Indeed, it did: He was also the "local" weather man in hundreds of radio markets across the nation, from New England to Honolulu.[19]

Now, *that* was interesting. All these trusting citizens heard the same familiar voice that we did. Yet, *their* Joe Zona was not *our* Joe Zona. Our Joe Zona actually experienced the same weather that we did. Reassuring as that might be, there is nothing local about weather prediction. Meteorologists in the U.S. develop their predictions based on three elaborate statistical models maintained by the National Weather Service. Each meteorologist may adjust or combine the outputs of the models, but she does so by analyzing complex data streams, not by looking out the window. This most physical aspect of our world, the weather, is better predicted by mathematical means than through direct observation. (WSI used to get calls in Bedford from a Hong Kong-based real estate magnate who would inquire periodically whether a particular day was propitious to sail his yacht across Hong Kong Bay.) So why *should* Zona be local—for Boston or any other market? The machines enabling weather prediction made location irrelevant; and the machines enabling the distribution of predictions facilitated displacement and substitution of personality, so that Zona could do the work of hundreds of local weathermen from his remote location.

Joe Zona's situation reveals the strange contradiction between rational fact and emotional meaning in machine-mediated interactions. An illusion of local presence for the vast majority of his listeners, he was also a higher quality (and lower cost) form of talent than any local station with a lean budget could source in local markets. For these reasons, the Zona model is hardly an isolated case. Sinclair Broadcast Group, which operates sixty-two TV stations across the country, has begun producing TV programs—"local" news, weather, and entertainment reports—at a central location and then shipping them to their owned and operated properties. Sinclair calls this "central-casting."[20] Clear Channel Communications, the nation's largest owner and operator of radio stations, pursues its own form of central-casting through nationally standardized program lineups; these appear on the airwaves as local programming across its more than 1,200 stations. Even the newscasts of TV network affiliates typically blend local production and central-casting, presenting local and national stories reported by on-air personalities whom viewers might readily assume are all local.

Such hybrid interfaces distribute human talent to generate far greater leverage than humans could on their own. For example, when Johnny Carson relinquished his thirty-year seat at NBC's *Tonight Show*, many fans experienced a sense of personal loss. After all, he "visited" millions of U.S. homes every night, going where "real" people could not go. Today, the appeal of late-night talk show hosts, such as David Letterman and Jay Leno, attests to the strong affective connections with viewers that human talent, through media, can forge over repeated exposures. Daytime-television personalities like Oprah Winfrey and Katie Couric are not literally our friends, but the hybrid interface of television—a machine enabled by people—makes us feel as if they are.

In this way, whether traditional or interactive, electronic media illustrate one aspect of the alchemical power of the hybrid interface wherein machines amplify human personality. This engine builds equity in consumer products or brands intimately associated with real or imagined human personalities. Marketers have long used the technique to forge personality-based emotional connections to sell otherwise unmemorable products such as grills (George Foreman), chicken parts (Frank Perdue), airline seats (Sir Richard Branson), and food processors (the Juiceman of Las Vegas). Like Joe Zona, each of these personalities becomes a synthetic point of personal attachment between a personality and a customer, and between a brand and a consumer. Each depends on technology to deliver the credible spark of humanity behind it. In a world that can bring human personality to life through automated or interactive interfaces online or via VRUs (in Tellme's original incarnation, you could play blackjack with an automated, if synthetic, Sean Connery over the phone), the opportunities are hardly limited to the media world. Before you dismiss the leverage model as unique to the media business, ask yourself, "How could key human actors have more leverage in my business?" The corporate world today could exploit such untapped leverage by deploying, for instance, an interactive representation of the CEO or a company spokesperson online or using kiosks. How else did Donald Trump and Sir Richard Branson build their personal and corporate brands? Their clever use of media and promotion to leverage their names and faces only hints at the potential. Consider what "The Donald" has gained through his prime time TV show, "The Apprentice," expanding his person and personality from book jackets, the business pages, and casino games to the participative interface known as reality television, even as his gaming business goes bankrupt.

An Integrative Example: First Direct

While the leverage model of machine-led hybrid interfaces may seem more familiar, the productivity model has also become mainstream. As we have noted, every call center representative actually operates a hybrid interface, where a machine (the telephone) substitutes for the physical presence of those who serve clients from centralized, remote locations. Using phone lines to "transport" customers to service workers, and not the other way around, increases service productivity. For example, a retailer could put tens of thousands of clerks in hundreds of retail stores or merely thousands of representatives in a call center. Or a brokerage or a bank could open a branch in every major city—or use a few call centers. That's what First Direct did to exploit the power of the hybrid-interface productivity model.

Headquartered in Leeds in the British midlands, First Direct began as a wholly owned subsidiary of Midland plc (now a unit of HSBC), then one of the Big Four national banks in the United Kingdom. Its genesis and separation from its parent company stemmed from in-depth customer studies commissioned by Midland. Midland never considered itself a market leader in terms of customer service. In fact, none of the so-called High Street banks pleased customers especially; after all, their U.K. oligopoly kept them viable even with marginal service quality. But, according to its customers, Midland was the *most hated bank* in England, fifth in a four-bank race according to one Midland manager. Still, Midland's researchers believed that they had found a small segment of extremely satisfied account holders among the more than 10 million extremely dissatisfied ones. Why were these few hundred thousand souls so satisfied? Because they never entered a branch.

A New Customer Segment

These individuals, who comprised a customer segment that would define the First Direct offering, had pieced together their own interface systems, combining telephone, ATM, and Royal Mail. Most had interacted with a branch manager early in their banking relationship, whom they could call when they had problems. Otherwise, they used available remote channels and never visited the premises. Predictably, this segment was younger, better educated, more technology-savvy, and concentrated disproportionately in professional jobs, with significantly better earning prospects than the general population.[21] These account holders did not maintain the highest bank balances, but they also did not require much attention. In terms of customer lifetime value, this segment was highly profitable for First Direct or any bank to serve.

Midland decided to base a new bank on the interaction model of this segment. Thus, First Direct was born. The new bank would operate from a single call center in Leeds.[22] It would do business entirely over the phone. When customers needed cash, they would use Midland ATMs and, later, those belonging to other banks. The bank would reinforce customers' visual impressions of its brand with a distinctive promotional campaign. Before opening up its phone lines, First Direct mounted a TV blitz. Chiat/Day in London created the ad campaign to be "disruptive" in agency parlance.[23] The central theme was that First Direct's banking proposition was not for everyone and that its approach to banking services would divide consumers into two segments: early adopters who embraced innovation, and laggards or Luddites who rejected it outright. First Direct's management had anticipated that it would have to educate the market about doing business with a bank that did not look like a bank, and bore no resemblance to the Big Four institutions. To introduce the concept, Chiat/Day mounted an early "roadblock" commercial, the technique of placing a TV advertising spot on several networks simultaneously, so that viewers could not easily avoid it by channel surfing. What purported to be a sixty-second spot for a newly released Audi Quattro sedan was interrupted by an apparent broadcast signal failure. In a matter-of-fact way, an attractive professional woman appearing in a new scene calmly announced, "This is a broadcast from your future." She said that she was speaking from 2010, at which time branchless banking had become commonplace. Around her, images of happy people floated through space with large shiny coins and the sound of laughter. The woman advised viewers to choose between an optimistic view of the future and a pessimistic one. The optimists, she said, should stay tuned to one channel, and the pessimists should flip immediately to another. Given the road-blocking placement across two channels, viewers received different instructions in each version. After these directions, the advertising message on each channel diverged. The optimistic one featured a British comedian wearing a white three-piece suit and top hat, dancing through the City of London on a crowded business day, and singing a humorous song about First Direct's wonderful branchless banking services. It concluded with First Direct's telephone numbers. The pessimistic one had the same comedian wearing a black suit and hat, wandering haplessly through the same streets, and wailing about the awful service at brick-and-mortar banks. It ended simply with "Thank You for Watching."[24]

The launch of First Direct was a resounding success. The bank reinforced its image among account holders through branded collateral materials, including distinctively packaged checkbooks, account ledgers, and file boxes to contain them—all jet black with large lower-case sans-serif white lettering. This stark clarity reinforced the bank's plain-speaking, no-nonsense approach to financial services and gave customers a sense of autonomy and mastery over their financial affairs, as opposed to the powerlessness and confusion that many customers experienced with traditional banks. In essence, First Direct launched a new kind of bank and, in the process, created a new lifestyle brand. By the early 1990s, it was the United Kingdom's fastest-growing bank and the only bank with significant brand equity. It had achieved customer satisfaction rates above 90 percent and nearly perfect levels of account retention. First Direct went on to qualify at the top of the U.K. banking industry's customer satisfaction rankings for twelve years, starting in 1991.[25] Only one of every two hundred customers submitted a complaint in 2002, and 74 percent of them appreciated how the bank handled their complaints.[26] The primary causes of account defection were moving out of the United Kingdom (the bank's only market) and death.

A New Breed of Banker

Not just its advertising set First Direct apart. The bank cultivated a new kind of banker, called the "banking representative" (BR), to deal with customers exclusively over the phone from its call center in Leeds. First Direct executives sought people "with the life skills that work well in our environment," individuals who were "used to juggling different demands and [who had] excellent communication skills," but had never worked in the Big Four banks.[27] Unlike the Big Four, First Direct recruited candidates from Leeds, not London; from redbrick universities, not Oxford and Cambridge; and from households, not offices. The recruits, mostly women taking a few years off from their careers as lawyers, accountants, and business managers to care for newborn children, needed jobs with flexible hours. First Direct required them to participate in extensive months-long training programs so that they would not only earn their banking licenses but also learn First Direct's unique approaches to customer relationship management. In Leeds and Yorkshire today, thirty-seven hundred BRs work from two warehouse-like call centers with attractive cafeterias, day-care facilities, and large open spaces filled with bright light and fresh air rather than cubicles.[28] The vibrant culture of these call centers differs dramatically from those of the large call center operations of other financial institutions. Moreover, First Direct's call centers are so decidedly relationship-focused that there is no visible evidence of such back-office functions as printing account statements or handling money; and the turnover rate among call center personnel is nearly one-third the U.K. average—12 percent as opposed to 30 percent.[29]

However, its homegrown bankers—even if they *were* smarter, kinder, more reliable, more solutions-oriented, and more empathetic—and its creative management went only so far to differentiate First Direct's services. The bank's customer channel was the telephone, but its capacity to serve customers effectively depended on providing each BR with a machine, namely a PC workstation. These workstations gave the frontline service workers access to the bank's customer information systems, which stored profiles and account information. These systems tracked three levels of information on account holders.

- First, they tracked customers' *identity* data—name, address, phone number, age, and income; how they came to the bank; and when they opened their account. An early application of caller-ID technology allowed a BR to see the caller's identity information on her PC the moment her phone rang, though BRs rarely greeted customers by name—too invasive—until the customers introduced themselves. Security information also appeared on this initial data screen. First Direct used a customer-friendly system of three or four rotating words, selected by customers. Banking representatives would ask customers, for example, for the third letter of a given password, which obviated the need for account holders to remember otherwise meaningless alphanumeric codes or their most recent deposits. No one has ever cracked this security system despite its fanciful character.

- First Direct also tracked histories of customer accounts, such as deposits, withdrawals, transfers, changes of job or address, and banking products or services. Though most banks viewed such histories as transactional, First Direct considered them *behavioral*, a source of insight into customers' future needs and desires. For example, an account holder who inquired about traveler's checks during one call might buy them on the next, and so the system would alert the next BR to an emergent cross-selling opportunity.

- The system tracked *emotional* data, such as a BR's observations of customer moods, personalities, and dispositions, which enabled other BRs to answer phones and interact with callers according to each customer's preferences and individual styles. In other words, at the start of each call, the system signaled not only *what* to discuss but also *how* to discuss it.[30]

Our research at First Direct revealed that this combination of customer profiling data, behavioral tracking, and insight into personality was a significant driver of the bank's high levels of customer satisfaction. When we asked customers to characterize their feelings toward First Direct, most would describe the experience as the most personal relationship that they had ever had with a financial institution—or, for that matter, with any large business. This finding sparked our curiosity in several respects. Obviously, First Direct provided no means of face-to-face interaction between employees and customers, and it acknowledged this abstract nature in several early TV spots. (In one TV commercial, an articulate, elegant man sat near a fireplace sipping tea; after a few seconds, he looked into the camera and, as if sharing a revelation, said, "I've never seen the people at First Direct, but I believe—*I believe!*—they exist."[31]) Also, unlike other direct-banking or brokerage firms, First Direct did not assign specific BRs to customers.

In fact, its capacity to provide live personal service on demand depended on routing incoming calls across hundreds of BRs regardless of who handled prior interactions. No customer would likely speak to the same BR twice during a typical multiyear banking relationship. Finally, First Direct offered no products or services beyond those of a traditional retail bank; its distinctiveness rested solely on its service delivery.

So why did First Direct's relationships feel so personal to account holders? The answer begins with the machine interface, the telephone, which enabled First Direct to centralize its operations in a few contiguous geographic locations. It staffed those locations with professionals hired for attitude, not skill, and trained to capitalize on those attitudes in retail banking. It confined its person-to-person interaction to phone lines so that it could utilize the unique intimacy of nonvisual media (as Joe Zona did through radio), and it gave account holders an array of consistently branded tangibles. By combining the emotional intelligence of its people with the machine intelligence of its front-office information systems, First Direct could lead its employees, as one executive expressed it, to "focus [not only] on sales or productivity, but also on the quality of the dialogue that they're having with the customer, [which] drives individuals' incentive payments at the end of the year."[32] The multifaceted insight into customers and their accounts established the quality of interaction with customers. In addition, expert systems immediately suggested not only what BRs should sell or cross-sell but also how they might meet customer expectations and anticipate future or emergent needs (as at Ritz-Carlton) most effectively. With the additional capacity to tailor styles of interaction—which depended critically on having the right human talent and the right machine systems—the bank matched its interactions on a segmented basis to customer personality types.

Could First Direct have established such a sense of intimacy in any other medium besides the phone? In face-to-face interactions, for example, BRs could not have integrated data from their PCs without losing eye contact with customers; they would sacrifice customer insight for customer connection. That need not happen over the phone. In this sense, First Direct selected the phone as the ideal machine interface for its purpose and arrayed its other interfaces to reinforce the phone. (First Direct launched online banking in November 1999, but only recently trialed a virtual agent named Cara who could field natural-language questions on-line.[33]) Still, there are challenges. While First Direct became profitable in 1995 and broke even on its initial investment in 1999, it has not grown far beyond a million accounts as some might have predicted several years ago.[34] In the global market, ING Direct has overtaken it by limiting its product lines to a few basic offerings (such as savings accounts with above-market rates), achieving scale through international expansion (operating in eight countries), and pursuing a particular segment of rate-sensitive customers in each country market (value shoppers), who have responded positively to flawless execution of basic direct-channel services—all of which keep costs low. Nonetheless, First Direct has consistently amassed the world's most satisfied and loyal banking customers by effectively deploying one of the most satisfying service interfaces in the financial services sector.

The Results

First Direct's operating statistics underscore what machines supported by people can do to enhance productivity (and leverage, too) of frontline workers who manage customer relationships, resulting in long-term competitive advantage. The productivity of First Direct's BRs dramatically exceeds that of traditional banking personnel; and over the past decade, that productivity gap has widened. In 1990, the bank's ratio of accounts under management to total employees was 367; at the Big Four banks, it was 139. By 1999, the gap had widened as these ratios had risen to 417 at First Direct and 161 at the Big Four. First Direct's customers have the greatest lifetime value in the retail banking industry, with 31 percent between ages 18 to 34, 50 percent between ages 35 and 54, and only 20 percent for 55 and over. While the Big Four banks compete through direct operations, not all offer 24-hour service, and they still operate through their parent bank's departmentalized organizations that require multiple specialists to deal with even standard customer requests. First Direct data indicate that more than half its customers call outside normal banking hours, and 90 percent of those calls are addressed and completed by the first person who answers the phone. Finally, by expanding its portfolio of services—from bill payment to car insurance to home loans and more—First Direct has capitalized on trust-based relationships to increase its customers' lifetime values.[35] It has aligned its approach to the market with the critical attributes that matter to customers. Even though First Direct does not operate ATM networks—its customers use HSBC's ATMs or those of the other Big Four—its customers are on average 15 percent more satisfied with their ATMs than other U.K. banking customers, including HSBC's.[36] Now that's a kind of alchemy at work.

First Direct showcases the four drivers discussed in previous chapters. The bank's use of the phone as its anchor interface renders its services *ubiquitous*; its distinct treatment of its related communications and collaterals reinforces this ubiquity, creating a compelling, consistent *physical* brand. Its use of databases and expert systems gives its machine-led hybrid interface distinct *cognitive* attributes, enabling effective execution of services, especially with the *intelligence* and *interactivity* of its banking representatives, who collect and use information to feed its databases and to interact appropriately with customers. Those representatives, selected for their attitudes rather than banking skills, forge *affective* bonds with customers through targeted personalization of interaction style and services. Finally, the cultural and systems-based *connectivity* among BRs within First Direct's operations enables any staffer to handle any call. Since customers can call at all hours on any day (including Christmas), the bank resembles an online service, but with live human warmth, empathy, and respect built into its primary interface.

Conclusion

One might argue that First Direct's interfaces are more complex than the archetypes explored earlier in this book. Indeed, the central hybrid interface that manages bank customer interactions is multilayered—machines (phones) answered by people (BRs) supported by machines (databases). This chapter explored how interface systems work and how to orchestrate them.

- *The hybrid interface archetype involves people supported by machines and machines supported by people.* Where neither people nor machines alone can do the job, managers and strategists must ask themselves, Which interface will put our company's best face forward to its customers? More often than not, the most efficient and effective outcomes involve hybrids, but these require managers to determine whether people or machine attributes will define the customer's experience of their interactions with a company and which should lead a given interaction.

- *Optimizing hybrid interfaces depends on ascertaining the optimal division of labor between people and machines in interactions with customers.* As we have seen, hybrid interfaces can create customer value that neither people nor machines alone can deliver. To combine these two elements of the work force, one must understand what people do best, what machines do best, and what people and machines do best together.

- *The two variants of hybrid interfaces generate different kinds of economic value with respect to human talent.* When machines enable people, people often produce more. Hence, call centers deliver sales and service more efficiently than workers who visit customers door to door. When people enable machines, people often gain leverage. Hence, media platforms such as the Web or the broadcast networks help personalities, endorsers, and celebrities to deliver their impact more effectively. Nonetheless, the themes of leverage and productivity describe the economic impacts of both people-led and machine-led hybrid interfaces.

- *There is synergy in hybrid interfaces.* Since people and machines in combination can bring such distinctive and valuable attributes to service interactions, their combined presence can often catalyze favorable customer impacts. For example, when a human service provider, informed by database systems, knows exactly what a customer wants before the customer can articulate her needs, the company exceeds customer expectations dramatically. A person without information could not have done so, nor could a machine without the reassuring human touch. Together, people and machines can interact with customers to increase the perceived value of companies and brands.

- *Ultimately, managers must determine the composition of any hybrid interface to optimize trade-offs between efficiency and effectiveness.* Machine automation in services (as in manufacturing) generally drives efficiency, and people delivering services generally increases effectiveness. As we have seen, however, the distinction is not so simple. For example, machines on the front lines can also drive effectiveness (by enabling personalization, enhanced customer privacy, and speed). By combining people and machines, managers can transcend the constraints of conventional thinking, driving top-line growth while compressing costs. But they can do so only if they understand what drives efficiency and effectiveness in human, machine, and hybrid interfaces, and how to combine and deploy such interfaces to optimize systems. Those capabilities are central to the new division of labor between people and machines.

In chapter 7, we focus specifically on the challenges of arraying simple and complex interfaces into optimal configurations of the attributes of people, machines, and hybrids.

7

MANAGING INTERFACE
SYSTEMS

JUST ABOUT EVERY COMPANY of scale has already deployed multiple interfaces that are costly and complex to operate. These
interfaces comprise systems that companies must maintain and
manage to minimize costs and satisfy customers. Countless new
interface forms are pervading even straightforward mass-market
retail businesses. Supermarkets, for example, with their thirty-five
thousand SKUs (stock keeping units) and their cashiers, service
agents, grocery baggers, and shelf stockers, have become hotbeds
of innovation in front-office automation, well beyond online ordering and delivery services such as Peapod and Fresh Direct. In
1999, only 6 percent of supermarkets had self-checkout lanes; in
2003, 38 percent did. One research group forecasts that 95 percent of supermarkets will use some form of self-checkout by 2006.
Machines such as U-Scan and Fast Lane, which cost $100,000 for
a four-lane installation, are quickly becoming ubiquitous consumer
standards in U.S. supermarkets.[1] Self-checkout lanes are only the
beginning. Stop & Shop Stores outside Boston use a device from

Symbol Technologies called "Shopping Buddy" that allows consumers to scan items when they put them in their shopping carts.
At the checkout, they simply pay what Shopping Buddy has tabu-
lated. Since the device tracks customers' selections and movement
in the store and maintains shopping records of previous store visits, it can deliver individually targeted promotions and discounts.
Shopping Buddy also enables customers to order prepared foods
remotely and pick them up from service counter personnel when
ready. SuperValu in Virginia installed kiosks next to deli counters
to perform similar functions, and Piggly Wiggly in South Car-
olina is using Pay By Touch systems that link fingerprint scanners
with customers' checking and credit card accounts so that customers can pay with a fingerprint. Other stores give customers on
premises wireless devices to notify them when prescriptions are
filled or photographs are printed.[2] In short, the retail industry has
transformed the store from its origins involving a few simple people-
dominant interfaces into a complex interface system involving an
array of people and machines.

However, supermarkets have developed essentially one-size-
fits-all service operations. Financial services pioneer Charles Schwab
recently reconfigured its existing interface system—which includes
financial services professionals at its three hundred Investor Centers nationwide; online resources for money management, investment research, and trading; financial advisors for high net worth
individuals through its U.S. Trust acquisition; and customer service representatives in its call centers—to meet the needs of several
major customer segments. As one senior Schwab executive commented, "many [customers] prefer live advice for recommendations,
but prefer the online channel for monitoring performance of their
accounts."[3] Schwab has modularized these interfaces and rational-
ized its front-office operations according to three major segments—
what it calls the independent investor who prefers machine-
dominant service interfaces online, the advised investor who favors
people-dominant interfaces offline, and the active trader who requires sophisticated machine interfaces online with occasional

access to human specialists through remote channels. For each segment, Schwab has configured an interface subsystem that targets the specific segment within its overall interface system. For example, when an advised investor goes online, the home page welcomes him with performance data for his overall investment strategy, whereas the active trader gets a home page with her trading dashboard and tools. Of course, these subsystems share interfaces—they all rely on Schwab's online site or its service lobbies—but "Schwab Personal Choice" attempts to anticipate and thereby optimize how these major segments employ its front-office resources.

Importantly, Schwab sells the same financial products and services across segments, but it lets customers segment themselves and then customizes *how* they access those products and services.[4] By preconfiguring interfaces and dialing up or down human contact based on segment preferences, Schwab limits the number of interfaces that any individual may encounter when accessing its services, thereby simplifying its operations and its customers' options. In complex categories, simplicity is a source of value. One American Express Financial Advisors advertisement recently observed, "You can't have too much money. But you can have too many retirement accounts."[5]

So how does a company maintain its brand character and effectiveness while automating its front office for efficiency? After all, new consumer-facing technology could depersonalize the brand experience. Worse, indifferent or unskilled frontline service workers could alienate customers altogether. Since interface systems can embrace many configurations, companies can harness the best attributes of both people and machines. The economics of mass-market grocery retailing will never again support the highly personalized service that small-town shopkeepers once provided, but machines like those at Stop & Shop can fill the gaps while trimming operating costs. Also, companies must carefully manage the allocation of people to customer interactions as Schwab did, because skilled and empathetic people are the scarcest, most non-scalable of an organization's resources. People must serve customers where their distinctly human capabilities and presence count most. Of course, many customers actually value the speed and convenience of machine interactions as long as they can access people when needed.

Unfortunately, too few executives focus on interface systems as sources of competitive advantage in contested markets, even when their companies lack any offering-based edge. Most deploy their service interfaces as loosely coordinated collections of customer touch points. At best, they function as themed portfolios of customer connections; at worst, they provide arbitrary points of customer interaction. Most companies have not yet considered their interfaces from a systems perspective, nor have they thought through the operational implications of multiple interfaces and interface subsystems. Similarly, they have often neglected such factors as customers' readiness to interact differently with companies, employees' preparedness to use new technologies, or technology's readiness to deliver new functionalities.

Interface proliferation, however, without a rigorous process that aligns interfaces with a company's strategy can increase business risks. If a company's interfaces work as an integrated system, then the business can realize a virtuous cycle of increasing revenues while reducing operating costs. If the interfaces work poorly, then they can confuse or repel customers and demoralize or alienate employees, spiraling into an adverse cycle that reduces revenues while escalating operating costs. In this sense, interface systems are a double-edged sword. This chapter introduces our methodology for integrating and optimizing a company's interfaces and their operation as a system. The chapter also goes inside the TV home-shopping retailer, QVC, a profitable epitome of interface systems thinking. In the next chapter, we give you a scorecard to assess how well your interface system is doing today—and how to improve it. Overall, we organize the integration and optimization process into five stages that we call the Five A's (see figure 7-1):

FIGURE 7-1

Five A's of Interface System Design

Assessment	Of the current experience
Aspiration	For customer interactions given customer desired experience
Alignment	Of front-office capabilities with the interface system
Articulation	Through deployment of interface and front-office configuration
Activation	Of customers and employees throughout interface evolution

- *Assessment.* What is the current customer experience of inter-actions with the company and its brand given its interfaces operating in the market today?

- *Aspiration.* What configuration of interfaces and interface systems will yield the desired customer interactions with the company and its brand(s)?

- *Alignment.* How must the organization's front-office capabil-ities, both human and machine, support the reconfigured interfaces and interface system?

- *Articulation.* Which plan of execution will deploy the desired interfaces and interface system and align front-office support?

- *Activation.* How should the interfaces and interface system evolve over time as customers, employees, and technology shape the interactions?

These five stages—assessment, aspiration, alignment, articula-tion, and activation—constitute the organizing framework for inte-grating and optimizing interface systems. It will help us to explain how a superior interface system has enabled QVC to so dramatically outperform its nearest competitor, Home Shopping Network (HSN), even though the two companies own and operate nearly identical consumer-direct business models. Both sell on television and online; operate call centers and distribution centers; reach nearly every cable-TV household in the United States; and run international operations in growth markets such as Germany and Japan. The Coke and Pepsi of their industry, they constitute most of the TV home-shopping market in the United States. They have based their sector of the retail industry on a reengineered front office and so, in the purest sense, they compete interface system to interface system. But in 2003, QVC generated $4.89 billion in revenues worldwide with EBITDA margins of 18 percent, while HSN generated $2.23 billion in revenues with EBITDA margins of 15 percent.[6] QVC realized sales of $580 from each active customer, while HSN generated only $383.[7] How can that be?

The QVC Story

As market leader in a $9 billion segment of the U.S. retail industry, QVC has grown rapidly since its inception in 1986. Its live broadcast features a continuous cycle of hosts presenting and selling products on air 24/7. With a 59 percent U.S. market share, QVC has risen from what was until recently a fragmented industry with nearly 50 home-shopping channels in the United States. Today, QVC reaches 86 million homes in the United States, 34 million in Germany, 13 million in the United Kingdom, and 11 million in Japan, and operates distribution and call centers in each market.[8] Industry estimates indicate that a cumulative 29 million people worldwide have shopped with QVC to date. Last year, its Web site accounted for over $500 million or roughly 15 percent of total revenues, making QVC one of the Internet's top general merchandise retailers. While QVC was not the industry pioneer—Lowell "Bud" Paxson established the category with HSN in the early 1980s—others have emulated its success formula around the world.[9] For example, LG Home Shopping, a chaebol-owned cable-TV network and one of the industry's many Korean players in South Korea, had sales in 2003 of nearly 2 trillion won (approximately $1.6 billion).[10]

Given its growth and scale, we might rank QVC among the world's most successful "electronic commerce" companies. Long before the Web, it proved how direct or technology-mediated retail channels and hybrid interfaces could produce extraordinary levels of customer satisfaction, loyalty, and operating efficiency. QVC touches its customers in diverse ways: For most of its customers, the shopping experience starts with broadcast programming and cascades through a variety of order-entry interfaces, including call centers, VRUs, and the Web site, with orders fulfilled by its own supply-chain operations and distribution centers. While an increasing proportion of new customers comes to QVC's interface system through the Web due to online marketing agreements with portals like AOL, most customers buy because of what they see on television.

Building the Better Shopping Experience

The idea of QVC—which stands for quality, value, and convenience—came to Philadelphia-based entrepreneur, Joseph Segel, while he was watching HSN in the mid-1980s. Segel, a master of direct marketing, made his fortune selling commemorative coins and keepsakes through The Franklin Mint. He immediately grasped the brilliance of using television as a direct-response medium to sell products, but deemed HSN's approach surprisingly unsophisticated. HSN's flaws spurred Segel to create a better version of TV home shopping that would tap the highest quality production values, high quality branded merchandise, and outstanding customer service. QVC went on air in November 1986. It completed its first year of operations with $112 million in revenues, making it one of the fastest-growth startups in history.[11] Its success demonstrated what Paxson had discovered with HSN: Combining television with compelling personalities, intriguing products, and remote ordering capability had an almost alchemical power among certain kinds of consumers to drive sales.

When QVC first aired, it projected integrity more than sophistication. Segel had opposed using professional models to display jewelry and apparel, and QVC's broadcast studio looked decidedly homespun. Until QVC moved to its current location in West Chester, Pennsylvania, in 1997, its TV studio and one of its call centers were located in the same facility so that hosts could see and hear telephones ringing as they talked about products on air. The main set was divided into quadrants and mounted on an enormous turntable. Each section resembled a different room in a home. At the top and bottom of each hour, when segueing from a jewelry show to, say, a cooking show, the set would rotate ninety degrees from the living room to the kitchen during the station break, and selling would resume in the new quadrant. In the early 1990s, QVC evolved from start-up to industry leader. QVC's acquisition of two competing businesses, Cable Value Network and J.C. Penney's home-shopping channel, added tens of millions of new viewers to its audience. Shortly thereafter, the media and entertainment executive Barry Diller, now chairman and CEO of InterActive-Corp., replaced Segal as QVC's CEO.

While Diller failed to use QVC as a consolidation platform to acquire CBS and Paramount Pictures, he did professionalize its broadcast production. He upgraded its on-air presentations, insisting on higher quality TV production, professional models, and higher-end brands, including a line designed by his future wife, Diane von Furstenberg.[12] Doug Briggs, QVC's current CEO, took over when Diller left in 1995. The network continued to improve operations. Today, QVC features some sixteen hundred products a week, two hundred fifty of which appear on air for the first time.[13] In 2002, it received thirteen thousand product pitches from entrepreneurs, with just 2 to 3 percent getting on air.[14] Meanwhile, the network's merchants have forged agreements with dozens of upscale brands, including Dell, Bose, Panasonic, IBM, and Pentax in consumer electronics; Timberland, Birkenstock, and Dooney &

Bourke in leather goods; KitchenAid, Braun, and Krups in small appliances; and Prescriptives, Philosophy, and L'Occitane in cosmetics. The network merchants must select products with sufficient intrinsic interest, complexity, or perceived risk for customers so that their presentation results in good television. Boring or simple products can flop, because QVC's hosts cannot as easily weave engaging stories around them for the six to ten minutes that an average product appears on air.

This story angle partly explains the network's appeal: QVC educates as well as entertains viewers. But another aspect of QVC's appeal derives from its ubiquity. The network reaches 96 percent of all cable and satellite homes in the United States. In any given thirty-second increment of prime time, about one hundred eighty-five thousand homes tune into QVC's broadcast; over the course of a typical prime-time hour, 1.5 million households view at least some portion of that hour, generating orders at a rate of five thousand to fifteen thousand an hour.[15] Most significantly, QVC has extraordinarily elevated levels of customer satisfaction and loyalty. Ninety-one percent of QVC customers rate its service in the top category of "excellent," and 93 percent of its annual domestic revenues come from customers who have purchased more than once that year. Of the seven million active QVC households in the United States that have purchased at least once from QVC in the past year, the average purchase frequency is ten times a year (for a total of fourteen items annually), and the average order size is $58.[16] Such loyalty is paradoxical, given that QVC operates in what is arguably the world's least trusted retail medium.

The QVC conundrum

How does a network that broadcasts in the land of schlock—among crudely produced home-shopping channels, and late-night hucksters selling abdominal rollers, salad shooters, Ginsu knives, and get-rich-quick schemes—distinguish itself as the nation's most trusted retailer? By doing more than retail. If you ask QVC executives, "What business is QVC in?" they struggle to answer, because QVC is a general merchandise retailer *and* an entertainment company *and* a direct-response marketer *and* an online merchant. The QVC effect on customers depends critically on its ability to captivate them, catalyze their purchasing behavior, and earn their trust over time.

Consider the network's customers. QVC sells to a relatively affluent audience, not to denizens of trailer parks as many believe. Its typical household has a median income of $61,300, which is 12 percent over the U.S. national average. The bulk of QVC's revenue comes from four segments: (1) affluent, suburban baby boomers in major urban markets, (2) affluent, suburban boomers in smaller cities, (3) single, college educated city dwellers, and (4) retirees with significant discretionary income. Customers are 90 percent female and concentrated in the major metropolitan areas of New York, Los Angeles, Chicago, San Francisco, and Philadelphia.[17] They are not cash-constrained but time-constrained, and they regard shopping as a sport or a form of entertainment. QVC's appeal clearly transcends the broadcast itself.

The broadcast signal stimulates the most purchases. The network presents six to ten products an hour, totaling some sixty thousand item presentations each year. It also showcases one item each day as Today's Special Value (TSV), which accounts for approximately 20 percent of QVC revenues. However, QVC's most precious asset is not merchandise but airtime. QVC must grow exclusively through the leverage and productivity of its interface system, unlike traditional brick-and-mortar or direct-to-consumer retailers that can build more stores or mail more catalogs. QVC is a finite storefront on television twenty-four hours a day, with each hour containing fifty-three minutes and thirty seconds of selling time, 364 days in a year (QVC goes off air on Christmas day).[18] So QVC's core business can drive growth only by continually increasing the sales productivity of its average minute of airtime—as it has, year over year, since its inception—or by establishing a second cable network. The choices regarding product selection and pricing (the merchandising function), on-air hosts and production values (the broadcast function), and aggregation and sequencing of products in any given hour of programming (the planning function) are critical to achieving the highest possible returns on each broadcast minute. These functions determine how it manages its interface system.

Leverage from personality

Since every minute of QVC programming is live (though shows may include B-roll or prepackaged video to show product demonstrations), nothing is scripted—no TelePrompTers. For every product presented, hosts have five lines of information and their own background research. The host typically occupies a set, either alone or with a product or brand representative, and directs her attention to one of three or four robotic cameras controlled by a line producer in a glass booth overlooking the studio. Hosts get some additional direction from their line producers through an earpiece. In that seemingly relaxed and natural setting, they sell at rates of hundreds of thousands of dollars an hour.[19] Today, all shows originate in QVC's new production facility at Studio Park, in West Chester, Pennsylvania, where the network built a sixty-thousand-square-foot studio—one of the largest digital TV production operations in the world—housing a twenty-five-thousand-square-foot cutaway suburban home that contains most of the sets.

The leverage on human talent is striking. Imagine asking the manager of a top department store—a Nordstrom or a Neiman Marcus—to introduce you to the top salesperson. If the manager obliges, then you'll likely meet a particularly polished and professional individual, perhaps an expert in men's suits or women's designer clothing. How much annual sales volume can this person achieve in such a store? The truly skilled might sell several million dollars of merchandise. At QVC, the average per capita sales productivity of any one host is *$200 million a year*, given QVC's nearly $4 billion in U.S. revenues and approximately twenty sales hosts on staff at any given time. The top hosts produce far more. Several years ago, one host was preeminent—Kathy Levine, who has moved to HSN. A former school teacher who auditioned in QVC's first casting call in 1986, she was instrumental in building the network's jewelry category. Today, jewelry represents approximately 30 percent of QVC's total programming, and the resulting revenue makes QVC one of the largest jewelry retailers in the world. Levine's individual talent could move several hundred million dollars of jewelry a year. Similarly, Bob Bowersox, the eleven-year host of "In the Kitchen with Bob," has sold about 900 tons of cookware at QVC. Both examples underscore the leverage on human talent that machine-led hybrid interfaces can provide.[20]

So how does QVC's interface system make so few people so productive in selling goods? One obvious answer is national reach. But HSN, whose signal reaches roughly the same number of households, is only half as productive in revenues. Another obvious answer is unique talent. But competitors can hire that away, as HSN did with Levine.

A Machine-Led Hybrid Interface in Action

A more complex answer relates to the productivity of the interface system itself. QVC interacts with customers through several interfaces more or less simultaneously, some of which are outbound (the broadcast), some inbound (the toll-free phone lines), and some both (the Web site). This configuration enables the network to glean real-time market intelligence over the phone lines and the Internet as it broadcasts programs. On the first of many visits to QVC, we witnessed this interface system in action one weekday in the late 1990s. The line producer that December morning was beginning a segment at the top of the hour featuring Kathy Levine as host. She was presenting thirty minutes of 14-karat-gold jewelry for women looking for something special to wear during Hanukah— talk about a niche market! At the start of the jewelry show, she introduced her guest, a bearded man dressed completely in black, who designed the jewelry line. With Levine's prompting, the designer began rhapsodizing about how versatile his designs were and how festive QVC shoppers would look wearing them during the holidays.

Levine asked him how someone might wear a certain gold necklace. The guest thought for a moment, then proclaimed that the necklace would look fabulous on a black angora sweater. Levine clearly liked this idea, and they discussed it a minute or two more. Back in the control room, we noticed that the line producer's screens began to exhibit a pattern once the guest mentioned the black angora sweater: The first screen displayed inbound call volume, and the numbers were rising rapidly. The next screen showed conversion of incoming calls to orders, and soon those numbers were rising, too. The final screen showed inventory levels of on-air items in the warehouses, and the numbers for the necklace began to decline. On the set, Levine was searching for new angles on the necklace, and so she asked her guest what other apparel might suit the piece. Again, he thought for a few seconds, then suggested a red angora sweater—for the Christmas season. That might not have appealed to the Jewish niche audience. On the screens in the control room, the sales pattern reversed itself. Call volumes began declining, conversions to purchase fell, and inventory levels plateaued. So the line producer barked into Levine's earpiece, "Kathy, get back to the black angora sweater!" Without missing a beat, Levine said, "Well, of course, red is nice, but black is *always* perfect with gold." Call volumes and conversions started rising again, and inventory levels began to fall.

Over the years, QVC has refined its use of real-time customer data and can check sales every six seconds. Since the merchandising mix has evolved to include more considered purchase items such as large screen televisions, this sales volume per minute information has led to occasional breaks from the six-to-ten items per hour format. A squad of PhDs on staff analyzes sales volume per minute throughout product presentations to determine the optimal length of time each product should remain on air in order to maximize segment revenue.

In terms of 2003 revenues, QVC was the number three TV network in the United States, behind only General Electric's NBC and Viacom's CBS.[21] But QVC behaves nothing like a traditional broadcast network. It is an interactive medium with an outbound channel for the broadcast signal and an inbound channel for audience response.[22] On a national mass market scale, QVC does what great salespeople always do when they make a pitch: They read client reactions, and they adjust their approach and behaviors to trigger a positive response. Where a salesperson can normally sell only one account at a time, QVC's hosts sell to hundreds of thousands of households simultaneously, adjusting what they do to match viewer segments that self-organize dynamically based on the products and brands on air. QVC does not buy Nielsen research, and its executives ignore standard TV metrics such as reach, frequency, and audience share. Rather, they determine and adjust their presentation strategies based on how many calls and Internet orders come in during each minute of broadcast time. Meanwhile, the network's capacity to take calls from viewers on air—so-called testimonial or T-calls—reinforces the credibility of the presentation, and helps create a sense of shared experience among QVC's viewing audience. QVC's customer service representatives often invite callers who have ordered an item before to discuss their experience on the air. When the line producer can slot the call into the show, the caller can talk to the people appearing on her television. She can talk about how much she loves a particular product on national television from her own home. In so doing, she provides the most effective risk-reduction mechanism known for prospective shoppers—namely, another customer's positive word-of-mouth delivered with mass-media efficiency.

The backyard fence

QVC reinforces its peer-to-peer or community dynamic among its viewers by ensuring that its hosts seldom appear to be selling. With rare exceptions, the network resists price markdowns as a strategy to move goods. Unlike its competitors, QVC sets a low introductory price for many items appearing on air for the first time; after initial exposures, the price reaches the level at which it will remain for the lifetime of that SKU. Other networks will lower prices to increase sales velocity, as their hosts admonish viewers to buy while supply lasts. When QVC hosts present products, they are interacting and educating. Darlene Daggett, QVC's president, has noted, "The advantage of QVC is sharing information that is relevant to the viewer, in a comfortable, conversational way." Hosts build relationships with viewers by providing a context that lowers perceived risk and stimulates consumer demand—by discussing histories of brands, demonstrating products, conversing with product spokespersons (the guests), and explaining complex subjects or categories. Hosts seldom hype price, and so they build intense personal followings among QVC viewers. These perceived relationships are crucial to QVC's success. According to QVC lore, the average shopper watches the network over a period of six months or a cumulative forty viewing hours before making a first purchase. While these numbers may be apocryphal, they point to a business truth: No one buys from strangers or without some element of trust—especially at a distance, through remote channels, from people whom they have never personally met. QVC creates a sense of intimacy with its viewers because it generates unique leverage from its hosts' personalities. By adjusting their behaviors based on viewer response, its hosts invariably interact with products and guests to reflect their shoppers' interests and connect with viewers through T-calls on air. In the long run, QVC gains nothing by selling products that viewers will ultimately return. Hard-sell tactics may briefly spike sales volume, but returns inevitably go up, customer satisfaction goes down, and loyal customers go away. Instead, QVC hosts explicitly say how they judge quality and value, whom certain products will suit, and whom they won't, under which circumstances, and on which occasions. QVC has cast this philosophy of host interaction as an operating principle called "the backyard fence."

The competitive advantage of engaging hosts who build relationships rather than push products increases tremendously when combined with the rest of QVC's interface system. QVC's broadcast screens reinforce the visual appeal. Like Google's home page or *The Wall Street Journal*'s front page, the QVC screen is instantly recognizable to channel surfers: it is consistently formatted, information rich, and easy to scan, with every fact that a shopper might need—like the product description, item number, retail value, QVC price, shipping cost, inventory level, call-in numbers, remaining airtime before the next product, and QVC's URL. This familiar context is one aspect of the backyard fence metaphor: A good neighbor is a trustworthy source of helpful information and dependable advice.

Underpromising and overdelivering

A good neighbor also keeps her promises. Most of the information on the QVC screen facilitates orders; each screen is akin to a page or an entry in a retail catalog that you might tear out to save for when you're ready to buy. When viewers wish to order, they can reach QVC in several ways. The traditional interface is the toll-free 800 telephone number. To handle call volume that has at times exceeded eight hundred twenty-one thousand calls in a single day, QVC employs roughly four thousand customer service reps who work in three shifts at four U.S. call centers. (In the United States alone, the network received over 150 million calls in 2003 and shipped over 97 million packages.[23]) To order from QVC, you must become a member by registering your name, address, and a form of payment, in exchange for a "Q-number." When customers contact QVC, the network processes their orders across its multiple interfaces, with almost 50 percent mediated by real phone representatives in call centers (supported by databases of customer profiles keyed to Q-numbers and product SKUs), 39 percent by VRUs, and 12 percent through the Web site. The telecommunications network operates with 99.99 percent reliability. Once orders are placed, customers benefit from QVC's second operating principle—"underpromise, overdeliver." While the network routinely promises that orders will arrive in seven to ten business days, orders frequently arrive in four or five days because the fulfillment operation ships 90 percent of orders within twenty-four hours—and 99.75 percent are shipped error free.[24] That rapid turnaround reflects QVC's adherence to a strict policy for items featured on air: QVC takes title to practically every SKU sold (except for bulky or perishable items such as exercise equipment or food products that manufacturers drop-ship), so that QVC can tightly control the quality of the fulfillment experience which goes beyond early or on-time delivery of orders to include distinctive QVC-designed packaging that reinforces in its trade dress the three core meanings of the brand.[25]

What's private stays private

Keeping promises builds trust. So does avoiding the hard sell. So does everyday stable and fair pricing. So does QVC's third operating principle of "strict confidentiality." While the direct-marketing field is notoriously rife with outbound telemarketing and e-mail promotions, QVC does almost none. From its inception, QVC has refused to sell customer data to third parties. In addition, it adheres to strict rules proscribing solicitation of customers, unless a service component accompanies the outbound message. Even when customers speak with call center representatives, QVC permits few cross-sells or up-sells, unless the up-sell relates specifically to an item that the customer is already buying. Even so, QVC will up-sell no more than 15 percent of callers a month and will up-sell no individual customer more than once a month.[26] Together, these three operating principles—the backyard fence; underpromise, overdeliver; and strict confidentiality—engender a high degree of trust between QVC's customers and its brand. In a recent survey using a seven-point scale measuring customer perceptions of "trustworthiness," 77 percent of QVC customers rated the network a 7. As we noted in chapter 3, such extreme ratings can reveal significant drivers of economic value. QVC is a case in point. Those customers who rated QVC a 7 were likely to repurchase at a rate 80 percent higher than those who rated it a 6.[27]

Interfaces and Linkages

Managing an interface system to produce such high levels of trust in a consumer direct channel is no mean feat. The interface system is only as strong as its weakest interface and its weakest link. If QVC's broadcast signal drives orders, but its call centers alienate customers by keeping them waiting, then the virtuous cycle disintegrates. If all of QVC's interfaces perform well, but the hosts and call center representatives promise delivery times that the distribution centers cannot fulfill, then the virtuous cycle also crumbles. This interdependence underscores an essential principle of interface systems design: Companies must optimize their interface systems along two dimensions—*separated* (performance of individual interfaces) and *related* (linkages among interfaces in the system). Striking the right balance is a subtle business, and even small changes to an interface system can have a huge impact. QVC learned this lesson the hard way. In May 2001, attempting to improve its on-screen look, the network introduced what management believed were more appealing and intuitive graphics and ran them for three weeks. Despite earlier testing, the new design proved too radical a change for viewers. Overall sales were down 20 percent. No matter what QVC did with its broadcasting, merchandising, and planning, it could not turn around these results. Finally, leadership discontinued the test and reverted to the status quo ante; with the old graphics back, sales immediately returned to normal levels. In this situation, nothing had changed among linkages within the interface system, but the context of one interface had temporarily weakened—and that weakness dragged the entire system down. This outcome is less surprising given how QVC generates "traffic." Word-of-mouth referrals draw 20 percent of QVC shoppers, but 70 percent discover QVC through channel surfing. To retain surfers, the network must reinforce consistent impressions among viewers over time—impressions of its always agreeable hosts, presentation style, and graphics—to reassure arriving viewers that they have returned to a context that they know and trust.[28]

Returns on relationship

QVC's interface system—with the virtuous effects of its optimized interfaces and linkages—also enables the company to enter new markets at lower cost. Remember that QVC shoppers invest a lot of time before making their first purchase. Given this unusual phenomenon, you might ask: With what or whom are they building the relationship—the network, the products, the brands, or the hosts? According to QVC executives, active customers do not buy within certain "consideration sets" of brands, categories of products, or price bands, as marketers might expect. QVC shoppers exhibit little such consistency and gravitate instead to specific hosts. In other words, they buy across diverse categories, brands, and price points—as long as they can purchase from people they trust. Of course, shoppers who are isolated, lonely, or overly dependent in their retail decision making might have reason to perceive these relationships as real. But QVC's entire interface system mediates real interactions and arguably builds real relationships. Call center representatives receive ninety-five hours of training before their first solo customer call; when they first interact with customers, they are both efficient service providers (the average call to place an order lasts less than two minutes) and effective personalities (they embody the friendliness, trust, and affective warmth that the hosts foster on television).

By tapping the leverage of viewer loyalty, QVC pioneered the PC category for TV home shopping in the mid-1990s. Aiming to broaden consumer awareness for Windows, Microsoft invited the network to become a launch partner for its new version of the operating system. QVC's management wanted to do business with Microsoft, but there was one problem: QVC did not sell PCs, and much of its largely female audience did not use them. To succeed, the network had to create the category among its shoppers in a hurry. Quickly, the network aired a show that paired a popular male host with a stereotypical PC company representative, aka a "propeller-head." The show flopped. The sale was just too intimidating. Subsequently, the network asked one of its popular female hosts, a relative technological novice, to assume the challenge. Building on her long-standing credibility with viewers, she began each of a series of PC shows on a set with a collection of unopened

PC boxes. A friendly and knowledgeable company guest joined her to discuss PCs in plain English. Each show began with the obvious questions: What is this thing? How do you set it up? How do you turn it on? What can it do? The audience empathized with the host's anxiety and moved up the learning curve with her. When she reached a point of comfort, her viewers bought PCs in droves. The Microsoft partnership proceeded, and QVC sold more units of Windows 95 in its first hour of airtime (12,200, to be exact) than Microsoft had projected for the first three weeks on the channel. That success established PCs as an important category in QVC's on-air merchandise mix—and heralded the network's single largest sales day ever. A few weeks before Christmas 2001, QVC featured several versions of a high-end Dell multimedia workstation as its TSV, with prices ranging from $1,700 to $2,200 when average PCs sold at retail for $700 to $1,000. In 24 hours, QVC sold thirty thousand Dell systems worth nearly $60 million, boosting the full day's sales to $80 million.[29]

The brand-building machine

The power of QVC to interact with viewers and build relationships has proven sufficiently robust that management now sees the network not only as a sales platform, but also as a brand-building machine. If every SKU gets an average of six to ten minutes of airtime on each appearance—not to mention endorsements from a host and actual consumers (in T-calls) as trusted third parties—then the value of QVC's offering in airtime alone to brand marketers is material. The audience reach of a minute of prime time on QVC roughly equals a $6,000 commercial placement on CNN's *Headline News*. By this measure, the venerable shoe company Birkenstock received over $7 million in "free" airtime in 2002 by appearing on the network, making its appealing spokesperson, Sally Combs, a celebrity among millions of QVC viewers. In recent years, QVC has begun investing airtime in its own proprietary or house brands. It launched a line of bedding called Northern Nights that is now a $100 million business annually (operated by just six professionals in QVC's merchandising department) and a cookware line called Cook's Essentials which in a few years has become one of the best-selling lines of cookware in the United States.[30] These proprietary brands attest to the strength of a well-orchestrated interface system and can create opportunities to grow sales through new interfaces. For instance, QVC'S $250 million Diamonique brand has performed well enough for QVC to test selling through a stand-alone kiosk in the Mall of America. In the same mall, the network has opened its own retail store under the QVC brand, to reach new potential audiences, to let customers experience the quality of a broad cross section of QVC's products firsthand, and to broadcast live from a proprietary and dynamic retail setting.

Why QVC Matters

In summary, QVC sells high-quality products, at fair prices, with truthful sales hosts who, in effect, offer neighborly advice across a backyard fence. Whatever the company does in delivering on its brand promise of quality, value, and convenience is reflected in its interface system, which then drives sales while building consumer trust. Since QVC has no physical presence for most consumers, since its customers buy few truly indispensable items, and since they can find roughly equivalent products at countless brick-and-mortar retail outlets, QVC must work harder to provide compelling reasons for its 7 million active shoppers to keep buying. And it has. By tapping hybrid service interfaces—largely machine-led—QVC has established unparalleled levels of consumer trust. In addition, QVC's amalgam of people and machines has helped it realize noteworthy levels of retail productivity. Just compare QVC to Wal-Mart and Sears: On a sales-per-employee basis, QVC achieves $444,455 in sales per capita compared to Wal-Mart's $170,886 and Sears' $165,157. On an EBITDA margin basis, QVC achieves $92,091 in margin per capita compared to Wal-Mart's $12,693 and Sears' $12,570.[31] Not surprisingly, Wall Street values Wal-Mart at a single multiple of revenues, Sears at a multiple of just one-quarter of revenues, and QVC at three times revenues. QVC's enterprise value was recently established at $14 billion in a transaction in mid-2003, when Liberty Media acquired the remaining 57 percent of QVC from Comcast for nearly $8 billion.[32]

Putting the Five A's to Work

As noted earlier, the Five A's model describes a structured process for rigorously evaluating the interfaces and interface system that a company currently deploys and then optimizing that system to drive gains in efficiency (cost savings) and effectiveness (performance). The objective is to accelerate growth in margin and volume by improving a company's management of its interactions and relationships with customers. This process comprises both interfaces and enabling front-office operations; it may result in cascaded effects on back-office operations, as well. While the process may appear largely focused on service strategies, the ultimate objectives are better enterprise economics and more robust competitive advantage, sustainable over time.

With these goals in mind, let's apply the Five A's to the case of QVC versus HSN, specifically their U.S. operations.* We selected the TV home-shopping sector because its very existence depends on a reengineered front office.[33] This approach to selling materialized to deliver retail services more efficiently and effectively to certain customers. Nearly all the resources of an organization like QVC go to front-office activities, because the company creates value for itself and its customers there. In this sense, the competition between QVC and HSN is truly a rivalry between one com-

---

*We based our analysis of HSN solely on our research team's data, which include viewing both channels during the same prime-time time slots in February and March 2004, information about brands and programming on QVC.com and HSN.com, customer comments from online bulletin boards, and secondary sources.

pany's interface system and another's. More interesting, they battle head to head in the United States with essentially equivalent scale and resources. For example, QVC reaches 86 million homes, and HSN reaches 81 million homes.[34] QVC has four U.S. distribution centers, and HSN has three.[35] QVC has three U.S. call centers, and HSN utilizes similar capacity, even though it outsources some of this activity to third parties in the United States and the Philippines.[36] The merchandise mix is comparable: 29 percent of QVC's programming incorporates jewelry; HSN has 23 percent. Both HSN and QVC log about 15 percent of revenues through their Web sites.[37] The customers are comparable, too: Like QVC, HSN's customer is typically female, average age 40, median income $63,000.[38] Both networks broadcast live twenty-four hours a day, though HSN operates a second channel under the brand name, America's Store, in some markets.

Despite these similarities in reach and resources, the divergence in revenue productivity is dramatic. QVC has nearly 7 million active customers while HSN has just over 4.5 million.[39] QVC realizes $48 in sales per active customer in an average month, while HSN only $32.[40] QVC generates a U.S. profit margin of over 23 percent, while HSN manages a margin of almost 17 percent on a top line that's roughly half the size of QVC's.[41] Why the performance gap between the two companies? The answer lies in the efficiency and effectiveness of their respective interface systems. Let's examine them through the lens of the Five A's.

Assessment: What is the current customer experience of interactions with the company and its brand given its interfaces operating in the market today?

We first examine the interfaces within the system on a separate and then related basis. (The next chapter details the assessment process. Here, for brevity, we take a less structured approach.) Like most companies, HSN deploys interfaces to manage each stage of the buying process, beginning with awareness and interest, then purchase, and ultimately post-purchase experience. Like

QVC, HSN builds awareness and interest primarily through its broadcast signal composed of on-air personalities, merchandise, and on-screen graphics. On any of these dimensions, HSN proves deficient. This is not to say that HSN lacks high quality hosts, merchandise, or production quality, but on the whole, QVC delivers a consistently higher level of quality in these aspects. Unlike QVC's consistently polished, friendly, and empathetic hosts, HSN's on-air talent is more variable—some hosts smile less, appear tense, and often push products. Rather than inform a customer's purchase decision, they often extol products with "fabulous!" or "incredible!" while admonishing viewers to rush to their phones before inventories run out. QVC's hosts build relationships with viewers and make products the center of attention, whereas HSN's hosts and guests sometimes appear at odds while discussing items.[42]

QVC hosts build a sense of community by asking viewers to call the dedicated T-call number on-screen; HSN solicits customer testimonials, but many times none call. In short, HSN hosts focus not on positive customer interactions but on managing down the network's inventory levels. In merchandising, QVC has a vastly superior, diverse stable of brands compared to HSN. Based on data available through their respective Web sites, QVC had 67 brands in electronics compared to HSN's 27; 126 brands in household and home improvement to HSN's 43; 72 brands in cooking and food to HSN's 22; 62 brands in beauty relative to HSN's 29; and 123 brands in fashion and accessories to HSN's 46.[43] In addition, QVC will regularly present six to ten new items in an hour, but HSN will sometimes dwell on three or four until inventory runs out. Furthermore, QVC carefully varies its weekly programming, but HSN often airs the same program in a three-to-five day time frame, sometimes airing the same segment (with slightly different products) eight times in one weekend.[44] This repetition becomes painfully obvious when guests are celebrities of yesteryear, such as Suzanne Somers and Susan Lucci. HSN.com bulletin board threads complain about the overexposure of such celebrities. Finally, regarding on-screen graphics, HSN provides only basic information on products, pricing, and item numbers. QVC presents this data but also rotates through seven or eight helpful phrases at the bottom of the screen, including its Web site URL, AOL keyword, various delivery options, and VRU call-in numbers. HSN engages in complex pricing, including coupons, special discounts, and charge cards, while QVC uses a simple, stable pricing structure.

At the purchase stage, HSN presents other challenges. When viewers call QVC, call center representatives will likely pick up the phone before the first ring and certainly by the second. When calls go to HSN, 65 percent are answered within twenty seconds—twenty long seconds during which 8 percent of callers hang up.[45] Some callers to HSN wait five minutes or more. QVC segregates callers according to those who prefer a real person and those who prefer automated ordering, offering a separate number for its VRUs. HSN funnels all callers through a VRU even if some want to reach a human being. Where QVC puts reps through ninety-five hours of rigorous training on the network's service values and business philosophy before they answer the phones and only utilizes in-house representatives, HSN outsources at least 25 percent of its call center operations, previously to Precision Response Corp, and more recently, to the Philippines.[46] In addition, it gives them incentives to aggressively cross-sell products not featured on air, compensating its representatives according to the incremental "offers" that result from add-on sales.[47] While QVC has improved its reps' efficiency from two minutes of talk time to place a customer order to just ninety seconds, and three minutes of talk time for a service call from four minutes, HSN encourages its reps to keep customers on the phone longer before concluding their orders. A manager at Siebel familiar with HSN's call center tactics describes, "Once you start buying, they're going to continue to sell."[48] Customers frequently object to such treatment, yet HSN has scripted its reps so that they have standard phrases to overcome objections. In contrast, QVC supports reps with intelligent systems entirely free of scripts, prompts, and function keys. Screens guide them through the order process, each showing tips for helping customers and options for how they could satisfy the person on the phone more.

Finally, regarding fulfillment, HSN offers inflexible shipping arrangements, with two options (five and ten days for delivery) as opposed to seven at QVC (from two to ten days with a variety of carriers). While QVC ships nine of ten orders in twenty-four hours, until recently HSN managed to ship only 87 percent within forty-eight hours—and shipments are often delayed.[49] Most of QVC's products are in its inventory, and so customers get what they order or are told that the item has sold out. HSN takes orders for what it does not have, then e-mails customers about putting them on a product "wait list." Returns are easy and quick at QVC, but difficult at HSN. According to customer comments posted on the Web, HSN phone reps lack knowledge of its merchandise or return processes, yet promise refunds that arrive late or never, or ultimately shift responsibility for problems to manufacturers.[50]

Are HSN's products inferior to QVC's, or its prices less favorable, or its hosts less compelling, or its broadcast production values lacking? No. But each interface in HSN's system suggests significant opportunities for enhancement. Senior management has evidently never subjected the related interfaces in the system to a process of integration and optimization. Consequently, similar hosts, merchandise, and broadcast production at HSN significantly underperform their potential in the market overall. In our diagnostic work, we often look for *pain points*, *choke points*, and *drop-off points*.

- When individual interfaces do not meet customers' needs, consumers experience pain. At HSN, such pain is rife, ranging from pressure tactics on broadcast to call center representatives who aggressively cross-sell.

- Any aspect of the buying process that prevents a customer from buying—that is, when on-screen graphics provide only partial ordering information or when callers wait extended times to order—is a choke point. At HSN, the wait times can be so long that viewers abandon their calls, and poor inventory tracking or flawed fulfillment results in shipping delays.

- Of course, significant pain points become drop-offs, but companies risk the greatest loss of customers moving across the linkages, not at the interfaces themselves. So you might conclude that HSN has deployed an interface system that prizes short-term transactions over long-term customer satisfaction. Yet HSN's profits depend on repeat customers, since the cost of acquiring new customers is often prohibitive. Hence, HSN's inferior financial results.

Aspiration: What configuration of interfaces and interface systems will yield the desired customer interactions with the company and its brand(s)?

In the aspiration stage, we ask managers to envision the ideal interface system in terms of the desired customer experience of interactions and relationships with their firm. This exercise typically starts by defining a company's brand for the company and customers alike. HSN's situation is more basic: The network is missing $2 billion from its top line, compared to QVC's. Consequently, HSN senior management might ask, "What customer experience could double revenue productivity for the hours of programming aired each day, week, and year?"

To generate viewer interest, HSN might address aspects of its broadcast, merchandise, and planning functions. In broadcast, more hosts with friendlier, empathetic demeanors, who connect with viewers rather than push products, would induce more loyalty. In merchandise, more national brands and product variety would spark the presentation of goods. In planning, a higher number of items in each hour—packaged with compelling themes—and less repetition of individual shows could lead viewers to make HSN "appointment television" or keep them watching. To help viewers gather information about potential purchases, hosts might focus less on talking viewers into buying, and more on *consistently* providing useful information that shoppers could use in their buying process. Such information might include detailed product descriptions, measurements, stories about how products were invented or manufactured, and ideas on how to use the products; hosts might also thoroughly examine and demonstrate every product they present every time, while soliciting similar information from their guests. To facilitate orders, HSN's broadcast could provide better guidance in many respects. Hosts might resist pressure tactics, which raise viewer anxiety about getting through to HSN in its often backed-up phone lines before products sell out. On-screen graphics could provide more than one phone number, especially when call center volumes peak; QVC, for example, provides numbers at all times for live reps and VRUs, while on-air hosts guide viewers to use less congested interfaces, like the Web site, when traffic is high. And service in the call centers could focus on ensuring positive interactions with callers rather than on incremental offers and revenues.

Finally, to ensure better customer interactions post-purchase, HSN could ensure that it keeps its promises to shoppers regarding shipping times, returns, and credits. The network might also shift its focus from a strict rules-based approach, which focuses on cost-containment, to one that maximizes customer lifetime value, thus easing its policies in favor of more positive customer interactions. (Like HSN, QVC has a thirty-day policy on returns, but it often accepts late returns up to ninety days or more—sometimes even several years.) Of course, to realize these goals, HSN would need to close the gap between reality and aspiration identified as pain points, choke points, and drop-off points in its interface system. By optimizing both interfaces and linkages, the network could better ensure that shoppers flowed smoothly through the buying process to minimize HSN's costs while maximizing its customer satisfaction. Based on these fixes to its interface system, HSN's leadership might then consider what distinctive positioning its brand could achieve vis-à-vis QVC—and how it might express that brand through its various interfaces to mediate interactions and relationships with customers.

Alignment: How must the organization's front-office capabilities, both human and machine, support the reconfigured interfaces and interface system?

In the alignment stage, we focus on changes required in front-office (and potentially back-office) activities to realize a company's aspirations. At HSN, the list of targeted transformations is potentially long. While HSN could achieve some obvious changes at relatively low cost—such as different on-air sales tactics, on-screen graphics, and call center incentives—many others require significant planning and investment. For example, QVC uses a unified database system to support its call center representatives when dealing with customers; reps can access all aspects of a caller's ordering process and all past orders through all interface types, while counting on features to automate outbound communications that advise of changes in order status or availability of desired products. In addition, QVC's systems have intuitive machine interfaces that integrate well with reps' conversations with customers. HSN could invest in such systems, but it would be unlikely to realize success unless it made other changes, such as running its own call centers, employing its own reps, and training them according to a philosophy of business that's about better interactions rather than quick-hit sales. Similarly, HSN could improve the post-purchase experience by investing significantly in fulfillment. Since these systems represent the final interface in the buying process, they have elevated impact on customer perceptions of value and experience. QVC has invested in a major supply-chain upgrade called the "glass pipeline." For HSN to do the same would require tens of millions of dollars. Of course, part of HSN's cost considerations inevitably focus on the underlying division of labor. Might HSN mediate more positive customer interactions at its various interfaces using people, machines, or hybrids? Might it produce lower overall systems costs by substituting or displacing certain front-office activities or functions? For example, in previous years, HSN generated a greater percentage of online purchases through its Web site than QVC, likely because its site allows shoppers to evade the pressure tactics of its call center reps. That might mean that improving the Web site—which has inferior search capabilities for product categories and purchase occasions—could generate positive business results at relatively low cost, as compared with more significant changes to people, culture, and organization, or technology systems.

Articulation: Which plan of execution will deploy the desired interfaces and interface system and align front-office support?

In execution, we advise managers to focus on this sequence of activities: *separate*, *relate*, and *integrate*. In the separate phase, companies should focus on reengineering individual interfaces to improve their performance in mediating customer interactions while lowering operational cost wherever feasible. In the relate phase, companies should entertain theories about how their systems might link to customers' buying processes. For example, both QVC and HSN assume that the broadcast signal is the *anchor interface*, putting the TV screen at the hub of the system. They also assume that customer traffic largely flows one way through the system—from the broadcast to the phone, VRU, or the online site, and then to fulfillment. For both companies, many shoppers start on the Web, consult the live broadcast (television, cable, or online), and order via the Web. Over time, Web sites will likely become anchor interfaces (as in South Korea among TV home-shopping operations), and buying processes will probably result in customers' following several standard flows through the system. As managers relate interfaces to one another, they must plan for current and future usage patterns—and facilitate linkages among interfaces according to current observations and future expectations—of their interface systems. In the integrate phase, managers must focus on meshing interfaces and linkages to maximize effectiveness. That requires closing performance gaps—pain points, choke points, and drop-off points—while amplifying operational efficiencies.

Activation: How should the interfaces and interface system evolve as customers, employees, and technology shape interactions?

During activation, managers should develop metrics that properly reflect their goals regarding desired customer interactions, relationships, and resultant customer experiences of their companies' brands or market positions. QVC illustrates that interfaces and interface systems can adapt positively in real time using metrics from vast consumer mass markets, resulting in opportunities for closed-loop learning. HSN illustrates that misguided metrics—such as call center incentives to drive sales rather than listen to customers—can distort interface systems to adapt negatively, despite customer dissatisfaction and defections. In this sense, metrics have enormous leverage on how interface systems evolve and adapt to changing patterns of customer usage as well as changing employee attitudes and behaviors in front-office jobs. In addition, metrics provide visibility for both customers and employees. QVC trains hosts, third-party guests, and call center reps in assorted programs through QVC University; the network continuously educates customers through hosts' on-air comments. Such integrative acculturation and orientation of employees and customers alike makes QVC's interface system work. In addition, training and measurement activities can also highlight opportunities for innovative or modified applications of technology, devices, or networks.

In effect, the activation stage is the acknowledgement that every interface system—especially when it involves talented people and smart machines—can become what technologists call a complex adaptive system, wherein the system self-optimizes. Such adaptation occurs through a combination of metrics and behavioral change, wherein feedback and training continuously drive better processes within the system. That has happened at QVC over the years and at other best practice companies discussed in this book.

HSN cannot easily emulate QVC's financial and operating results because of how QVC fine-tunes its interfaces and interface system year after year—based on appropriate metrics, incentives, and management systems and yielding a $2-billion revenue advantage.

CONCLUSION

As companies today compete increasingly on the quality of their interactions and relationships with customers, strategy must inevitably focus on interface systems. As QVC illustrates, back-office activities still matter—someone must originate, design, and manufacture the products that QVC sells—but companies often win or lose in contested markets according to their front-office performance. Most businesses are competing as QVC does with HSN—interface system to interface system. Put simply, interface systems are a company's ultimate expression not only of its brand but also of its strategy. This chapter covered several themes of the book:

- *Interface systems exist to mediate interactions and relationships with customers to maximize a company's efficiency and effectiveness.* The building blocks of such systems are interfaces composed of people, machines, and hybrids. At each interface, managers must prudently divide labor between people and machines, increasing value for customers while decreasing costs for their firms. When configured appropriately, interface systems can drive a company's leverage from human talent and productivity from overall operations.

- *Optimizing interface systems occurs in three stages—separate, relate, and integrate.* Each separate interface must optimally balance the delivery of customer value and the management of company costs. Interfaces must link or relate to one another according to how customers interact with them in the buying process and how companies facilitate these customer interactions. And companies must integrate interfaces into

systems that minimize or eliminate performance gaps
referred to as pain points, choke points, and drop-off points.

- *In deploying interface systems, management must determine the
  anchor interface(s) and its theory of customer flow(s).* In every
  interface system, some interfaces matter more than others.
  Customers may use one with great frequency but use another
  one for longer durations. These interfaces must receive management's highest priority for attention and resources. Similarly, in every interface system, customers can flow down one
  or more logical paths, based on how customers interact in the
  relevant buying process(es). A company cannot optimize an
  interface system without a point of view about anchor interfaces and customer flows.

- *The Five A's do not describe a linear, sequential, or finite process;
  rather, they describe a perpetual and related cluster of management
  activities.* These activities are intrinsic to managing and
  optimizing interface systems for competitive advantage.
  Since the best interface systems are complex adaptive systems,
  managers should not apply the Five A's in any prescribed
  sequence. To sustain competitive advantage, managers must
  continually measure their interface systems against customer
  and employee attitudes and behaviors as well as technology
  trends—for the life of the system. After all, the execution of
  strategy is the task of management—and interface systems
  are the ultimate expression of strategy.

8

The Interface Audit

A s the foregoing discussion makes clear, designing and deploying interface systems is a complex managerial challenge, involving the conception, implementation, and optimization of individual interfaces as well as the integration of those interfaces into coherent systems. To accomplish these tasks, managers must coordinate people, processes, and technology systems. Successful execution depends not only on traditional business disciplines such as marketing, service management, human resources, and IT, but also on less mainstream fields such as psychology, usability, aesthetics, and industrial design. It is not possible in these pages to furnish an exhaustive review of our methodology for creating and optimizing interface systems. A complete front-office reengineering effort requires analysis of opportunities at each potential service interface for *substitution* (deploying machines instead of people), *complementarity* (deploying hybrids of machines and people), and *displacement* (using networks to shift physical locations of people and machines). In this chapter, we provide an approach for assessing a company's current interface system using a rigorous set of analyses integrated in a scorecard.

This assessment serves as a starting point for managers who are considering the broader question of what actions they may

Eleanor J. Kyung, in collaboration with the authors, developed the methodology used in this chapter.

take to establish or sustain competitive advantage using their company's interface system. In this regard, the tools presented in this chapter represent a diagnostic for the interfaces a company currently deploys to determine (1) if the company is adequately managing and optimizing its interfaces both separately and together as a system, and (2) how those interfaces perform relative to competition. This assessment consists of two parts:

1. *An inventory*—What separate interfaces has your company deployed and how effectively do they interconnect, or relate, to one another as a system?

2. *A scorecard*—What performance ratings do your company's interfaces merit relative to competition on key success factors of interface systems?

The inventory is a tool to help you understand the entirety of your interfaces and interface system—what interfaces your customers use to interact with your company, how you might classify those interfaces according to interface archetypes, how well the linkages within your interface system work, and where decision-making authority lies. By outlining those components of your interfaces and interface system, you can quickly assess whether your organization functions with an integrated face to customers or as a collection of disaggregated or loosely coordinated corporate faces. Once you have completed this exercise, the scorecard provides a framework for evaluating the interfaces you have in place along critical performance attributes of superior interface systems: (1) effectiveness, (2) efficiency, (3) consistency, and (4) adaptability. As a point of clarification, this scorecard is not meant as a report card, assigning an arbitrary letter grade to performance, but rather as a mechanism for examining whether your company is reaping the benefits of systems thinking relative to competition. To that end, analyzing and understanding the gaps between current practice and future possibility is more important than achieving some absolute or relative score.

Depending on your level within your organization, you may wish to analyze your company's interfaces and interface system from a functional, business unit, or enterprise perspective. The principles remain the same, although the complexity and associated opportunity for impact will likely rise with your level in the organization. You will derive the greatest value from this assessment by keeping in mind a specific customer segment (or multiple segments) cross-tabbed with specific usage occasions. As you evaluate your company's performance, make sure you are considering *all* interfaces that customers might come in contact with, rather than those managed, for instance, by a single business unit.

INTERFACE SYSTEM INVENTORY

The best way to understand your interface system is through visualization, even if that visualization might appear simplistic. The following four steps will provide you with an initial interface system inventory. Again, consider at least one key customer segment as you proceed; you may wish to review figure 8-1, which provides a sample template for tackling Steps 1 through 3.

Step 1: Make a list of all of your customer-facing service interfaces—people, machine, and hybrid. For a visual representation of each interface archetype, indicate people-dominant interfaces with a white circle, machine-dominant interfaces with a gray circle, and hybrids with white and gray concentric circles (depending on whether it's a people-led or machine-led hybrid interface). For more complex interfaces involving multiple layers of people and machines, you may wish to add additional concentric rings to indicate important enabling interface elements.

Step 2: For each interface indicated, take note of the business units responsible for its deployment; the activities, processes, and systems associated with its support; and any individuals or departments with key decision-making authority.

FIGURE 8-1

Example: Interface System Template (Conceptual)

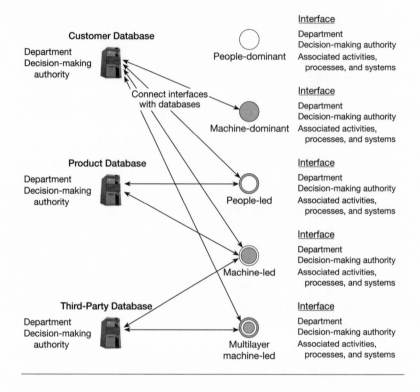

Step 3: Create an inventory of your company's information systems as they pertain to customer information, including profiling on demographic, psychographic, transactional, and emotional dimensions. Also, document your company's information systems as they pertain to product and service offerings, including any available customer assessments (e.g., peer reviews or peer-to-peer communications) regarding those offerings. Finally, create a list of third-party data resources that your company does or could use to complete its information resources. As you compose this picture, take note of the business or organizational units that house the various data resources, if they interconnect (and to what extent), and the key individuals with decision-making authority.

Step 4: Draw the linkages among the interfaces within your interface system and the information resources to which they are connected.

Step 5: Draw connections among interfaces within your interface system indicating which interfaces link to others. For example, do single-purpose kiosks direct users to service counters when necessary? Do call center representatives direct customers seeking detailed information to relevant Web pages? Do account transactions in a retail lobby appear in real time in customer data displayed by ATMs or on the Web? For a sample diagram, see figure 8-2, which illustrates

FIGURE 8-2

Example: QVC Interface System (Conceptual)

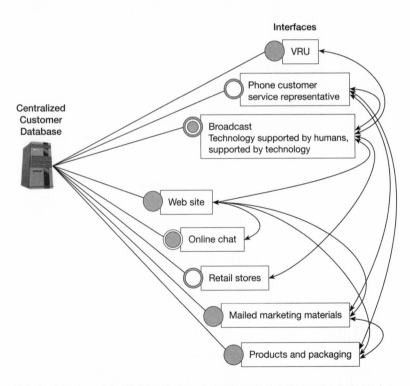

Note: For the purpose of simplified classification, we consider product, packaging, and physical market communications machine-dominant interfaces.

Inventory: Key Questions

Question	Answer
1. Evaluate each of the interfaces in your company's interface system on a scale of 1 to 7 according to what extent each fulfills customer needs and expectations (average of responses below).	1 2 3 4 5 6 7
A rating of 7 indicates that an interface has exceeded customer needs and expectations by a wide margin. A rating of 1 indicates that an interface falls significantly short of customer needs and expectations. A rating of 4 indicates that an interface has met but not exceeded customer needs and expectations.	
Frontline service workers	1 2 3 4 5 6 7
Call center representatives	1 2 3 4 5 6 7
VRUs	1 2 3 4 5 6 7
Web site	1 2 3 4 5 6 7
Online chat	1 2 3 4 5 6 7
E-mail	1 2 3 4 5 6 7
(Extend list to include additional interfaces)	1 2 3 4 5 6 7
2. Examining your interface system through the lens of a specific customer segment, is there a person, group of people, or business unit ultimately responsible for coordinating interactions and relationships with this segment?	Yes No
While it's not essential for one person or organization to be ultimately responsible for coordinating interactions and relationships with a specific segment, every company needs clarity regarding the manner in which relevant interfaces and related business systems will engage with specific segments. Configuration of interfaces by segment should reflect explicit, integrated, and coherent choices.	

QVC's interface system in line with our discussion of QVC in chapter 7.

As you step back and examine your rendering, use the diagram of your interface system to answer the following questions in figure 8-3. The first question focuses on evaluating the efficacy of individual interfaces. This question includes criteria against which to assess each of the interfaces in the interface system based on how effectively they meet or exceed—or, in the worst case, fall short of—customer needs and expectations. This data enables an assessment of opportunities for upgrading each interface in the system on a case-by-case or *separate* basis. The remaining ques-

FIGURE 8-3 (continued)

Question	Answer
3. Examining the connection of interfaces and database systems, is the relevant information for each customer segment integrated and updated in real time? Are changes resulting from interaction through one interface immediately reflected in changes to others?	Yes No
Data consistency is an essential underpinning across a well-orchestrated interface system. Ideally, information entered and retrieved from key interfaces—call centers, automated phone systems, Web sites, retail stores—should be reflected immediately in changes across all interfaces. Legacy systems may not always make data integration possible in the short term, but an integrated or centralized system is essential to customize interactions for individual customers. A leading source of customer frustration results from information known to one interface not immediately accessible through other interfaces.	
4. Examining the connection among customer-facing interfaces, does every interface appropriately point to the other interfaces in the system?	Yes No
Reaping the benefits of an interface system requires that each interface reflect linkages to other interfaces whenever appropriate, such that interfaces can optimize customers' interactions with the system. For example, an interface could refer customers waiting in a phone queue to a Web site or a VRU. Smart systems actively manage customer interactions, matching usage occasions with appropriate interfaces. Disconnects can occur when key interfaces fail to reflect system knowledge.	
5. Examining your inventory of interfaces, is there a mix of interface archetypes available to customers that matches interfaces they already use in daily life?	Yes No
While the appropriate interface mix depends on the profile of your customer segment, an interface system, by definition, requires a range of interfaces to mediate customer interactions and relationships efficiently and effectively. To what extent does your interface system utilize interfaces with which each customer segment is already comfortable, familiar, and literate?	

tions focus on evaluating the efficacy with which your company's interfaces function as a system or on a *related* basis. Together, these questions will support you in determining the degree to which your organization has taken a strategic stance toward effective management of its interfaces and interface system as sources of competitive advantage.

For Question 1, it's important to take note of those interfaces that do not merit a score equal to or greater than 4. These are interfaces that are failing your company's customers and, likely, its employees. These interfaces demand management attention to improve them each on a stand-alone basis. Individual interfaces must at minimum meet fundamental customer needs and expectations as a hygienic requirement. Otherwise, a company risks building an interface system composed of worrisomely substandard elements. Such fixes should ideally occur before significant investment of resources in optimizing integration of interfaces in the context of a broader interface system. As we have noted, customers judge a company based on a composite set of impressions formed through interactions with many, if not all, of a company's interfaces. As a result, an interface system is no stronger than its weakest interface. Linkages among interfaces are important, but the first priority of management must be to ensure that each interface performs at adequate levels of efficacy from a customer perspective. To evaluate how your interfaces function as a system, return to Questions 2 through 5. If you answered yes to Question 2, your organization has already restructured its managerial decision making and accountability around customer interactions and relationships—not just around business units or functions. If you answered yes to Question 5 but to none of the others, then your system may be deployed but not actively managed. If you answered yes to Questions 3 or 4, then your organization has begun to adopt interface systems thinking at a process level, actively instituting human resources, technology, and information systems to facilitate the deployment of a coherent and integrated corporate face to the customer.

If you answered yes to all of the questions, then your company is strategically managing its interface system. However, there are few organizations today in which managers can definitively answer yes to all four questions. Even so, every organization must actively track customer behavior for opportunities to innovate with respect to new front-office configurations and new interface technologies. In other words, competitive advantage through interface systems accrues when organizations are configured to adapt to customer behavior and subsequently alter interface systems as needed. Such adaptation doesn't require a company to adopt cutting-edge technology as soon as it becomes available. Rather, companies should actively monitor available interface technologies and choose carefully among those that might strengthen their competitive position. Managing technology requires a balancing act between emergent interface capabilities and customer and employee readiness to adapt to new technologies and processes.

Where you answered no to any one of the four questions, you should catalog the hurdles that prevent your organization from reaching its goals. Now that you have determined the relative quality of your company's interfaces and to what degree they are functioning as an interface system, you can evaluate the larger question of your interface system's contribution to your company's competitive advantage. To this, we turn to the Interface System Scorecard.

INTERFACE SYSTEM SCORECARD

Throughout this chapter, we have discussed your company's interfaces according to their archetypes and the underlying business processes enabling your interface system. There are four characteristics that apply to an optimally configured interface system:

- *Effectiveness*—Does your interface system enable your company to manage customer interactions and relationships with maximum effectiveness?

- *Efficiency*—Does your interface system enable your company to manage customer interactions and relationships with maximum efficiency?

- *Consistency*—Does your interface system enable your company to provide scalable and consistent service across the entire organization?

- *Adaptability*—Does your interface system possess the capability to integrate new interfaces and configurations based on evolving customer needs?

Evaluating an interface system's strengths according to these four characteristics forces us to broaden our consideration beyond actual interfaces deployed, including those enabling elements of front-office and back-office activities that support the organization's

FIGURE 8-4

Characteristics for Interface Advantage

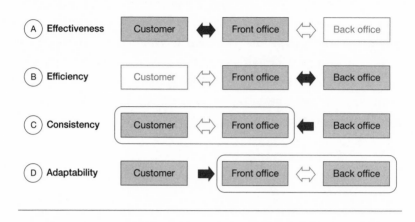

customer interactions and relationships. Figure 8-4 illustrates how these enabling elements relate to the four interface system characteristics.

We evaluate effectiveness based on whether and to what extent the front office is delivering high-quality interactions and relationships to the customer (see Figure 8-4: A). We evaluate efficiency based on whether and to what extent the front office and back office are functioning in concert to support mediation of interactions and relationships at lowest cost (see B). We evaluate consistency based on whether and to what extent the back office enables consistently high-quality interactions between the customer and the front office (see C). And we evaluate adaptability according to whether and to what extent the front office and back office adapt together to incorporate changes in customer behavior, particularly as customer needs, frontline work practices, and interface technologies evolve (see D).

The scorecard consists of a series of metrics to evaluate an interface system's performance. The sections on effectiveness and efficiency describe the system's *execution* of relationship delivery. The sections on consistency and adaptability describe the system's *structure* for relationship delivery. These characteristics are scored on a scale of 1 to 7 and measure your company compared to industry benchmarks. Note that we have designed these questions to fit most business sectors, but you may need to modify some questions based on the nature of your company and its industry.

As your organization and its interface system evolve, consider these questions against benchmarks outside your industry—including any companies with which your customers may come into contact. Customers compare experiences across sectors all the time—for example, between sales personnel at Nordstrom and tellers at the local bank branch or between transactions on eBay and transactions using the ATM. Especially as you find your organization scoring higher within its industry, it's critical to broaden the competitive set to include other industries and organizations, particularly those that can mediate interactions and relationships in ways that your customers value.

Effectiveness

Since effectiveness refers to the quality of a company's interactions and relationships with customers, it should be measured initially from the customer's perspective. In this sense, the characteristic of effectiveness measures how successfully a company's interface system—enabled by front office and back office—meets or exceeds customer expectations of interactions with the company. From the company's perspective, it is important to measure the impact of successful mediation of customer interactions and relationships in terms of average order size, frequency of purchase, size of customer base, customer acquisition, and customer retention. But these business metrics are by-products of getting interactions and relationships right for customers. As context, it is helpful to track your customers' attitudes regarding their relationships with your company over time, which you can determine through any of a number of research techniques. For example, a leading indicator of loyalty is a customer's propensity to recommend your company or its products to others.[1] Metrics such as these support a company in monitoring its performance, but they do not pinpoint specific managerial actions.

According to our research, several key factors directly influence the quality of customer interactions and relationships with companies. First, customers' perceptions of interactions must demonstrate that the company is *reliable* at every interface, *consistent* across interfaces, and capable of providing intrinsic *value* at each point of contact. Beyond these core elements, memorable and distinctive interactions depend on a company's interfaces exhibiting some combination of the following attributes:

- *Accessible:* Interfaces, whether people, machine, or hybrid, are easy and convenient to access at all relevant places and times.

- *Intuitive:* Processes associated with interactions at any one interface are simple and straightforward and take no longer than necessary to execute.

- *Aesthetics:* Interfaces are agreeable, appealing, or even compelling from a visual or broader sensory perspective.

- *Personalized:* Behaviors of interfaces are personalized by individual customer or customer segment, usage occasion, and type of interaction.

- *Balanced:* Interfaces integrate appropriate people and machine capabilities to deliver optimal interactions and, when problems arise, manage service recovery.

- *Pleasurable:* Interfaces mediate interactions in ways that customers find pleasing in distinctive or differentiating ways.

We evaluate the effectiveness characteristic from the perspectives of customer perceptions of interaction and relationship quality as well as performance metrics tracking customer value over time. With these points in mind, you may wish to turn to the questions in figure 8-5 to determine your company's performance with respect to effectiveness metrics.

Efficiency

A central theme throughout this book underscores the role of technology in enabling companies to mediate interactions and re-

FIGURE 8-5

Interface System Scorecard: Effectiveness Metrics

Rate the performance of your company relative to its industry along the following metrics using a scale of 1 to 7, where 7 indicates that your company is a clear industry leader, 4 indicates that your company is on par with the industry, and 1 indicates that your company lags significantly behind the industry.

Effectiveness Metrics	Score
1. Critical customer metrics (average of responses)	1 2 3 4 5 6 7
Average order size	1 2 3 4 5 6 7
Average order frequency	1 2 3 4 5 6 7
Size of customer base	1 2 3 4 5 6 7
Overall retention rate	1 2 3 4 5 6 7
Proportion of new customers	1 2 3 4 5 6 7
2. Interactions—performance versus expectations (average of responses)	1 2 3 4 5 6 7
To what extent do customers perceive your interfaces and system as:	
Consistent	1 2 3 4 5 6 7
Reliable	1 2 3 4 5 6 7
Useful	1 2 3 4 5 6 7
3. Customer perception of interactions—distinctive versus generic (average of responses)	1 2 3 4 5 6 7
Evaluate the ability of your interface types to provide distinctive customer experiences along the qualities outlined on page 218 by interface:	
People-dominant interfaces	1 2 3 4 5 6 7
Machine-dominant interfaces	1 2 3 4 5 6 7
People-led hybrid interfaces	1 2 3 4 5 6 7
Machine-led hybrid interfaces	1 2 3 4 5 6 7

lationships with customers not only with greater effectiveness but also at lower cost. As a result, efficiency metrics, in conceptual terms, track a company's cost per "unit" of customer interaction or relationship effectively mediated or delivered.

To calculate efficiency metrics, you must first determine the approximate costs of interacting with customers at each interface across your organization's interface system. In effect, you are determining the marginal cost of doing business at each interface. Fill in a simple grid such as table 8-1. The data in this table enables you to consider interface efficiency in light of what you already have assessed with respect to interface effectiveness. Using this approach, you may identify opportunities to migrate customers from

TABLE 8-1

Interface Cost per Transaction Grid

	INTERFACE				
	Retail Store	Call Center	VRU	Online	Automated E-mail
Cost per Transaction	$	$	$	$	$

higher to lower cost interfaces without diminishing the effectiveness of interactions for customers. You may also find opportunities to increase effectiveness while lowering costs. (For example, speed, accuracy, and anonymity are attributes of interactions mediated by more efficient machine interfaces, *and* they are attributes that customers on some occasions will prize highly.)

To develop a more rigorous picture of how customers use your company's interface system, it is important to catalog the distribution or flows of their interactions across interfaces in the system. Fill out a grid similar to table 8-2, again with one or more important customer segments in mind. In the first column, list the most common usage occasions for customers as they interact with your company. In the second column, indicate the number of customer interactions by usage occasion in a standard time period that makes sense for your business. In the remaining columns, fill in each row according to the distribution of interactions across interfaces accessed by customers for each usage occasion. For example, searches for product information might be 50 percent in store, 40 percent online, 6 percent by phone, and 4 percent by VRU and e-mail. To aid analysis later in this exercise, sequence the columns for interface types from left to right in order of decreasing cost per transaction.

Given the range of customer profiles, segments, and usage occasions—and their implications for how customers may wish to interact with your company—you can now make determinations regarding opportunities to migrate customers to lower-cost interfaces. In cases where you are employing people to handle rote transactions, for example, the attractiveness of migrating cus-

TABLE 8-2

Interface–Usage Occasion Grid

Usage Occasion	Total Interactions/Period	% OF CUSTOMERS USING EACH INTERFACE					
		Retail Store	Call Center	VRU	Online	Automated E-mail
1. Search for Product Information	#	%	%	%	%	%
2. Search for Store Information	#	%	%	%	%	%
3. Purchase	#	%	%	%	%	%
4. Product or Service Support	#	%	%	%	%	%
5. Address Change	#	%	%	%	%	%
6. Request Information from Sales Rep	#	%	%	%	%	%
7.	#	%	%	%	%	%
8.	#	%	%	%	%	%
9.	#	%	%	%	%	%
10.	#	%	%	%	%	%

tomers to automated interfaces will become obvious. In other cases, migration may not prove as easy or as obvious as a means to increase efficiency and effectiveness simultaneously. This is where you may identify opportunities to introduce more complex interface types—such as hybrids—into your interface system in place of, or in addition to, high-cost people-dominant alternatives. Once you have completed the grid, answer the questions evaluating the efficiency of your company's interface system in figure 8-6.

It should go without saying that collecting data of this kind can prove challenging operationally, and sometimes politically complex, because success in data gathering depends on access to performance information from disparate parts of the organization. In organizations that have not historically focused on the quality of their interactions with customers, the data may not exist at all. In that case, a lack of data in and of itself serves as a useful diagnostic, though it does not necessarily advance the cause. In any event, the mere act of collecting such data, if managed with savvy and sensitivity, can serve to heighten awareness across an organization of the strategic importance of quality in customer interaction and relationship management.

Consistency

Consistency characteristics relate to both efficiency and effectiveness. A well-orchestrated interface system mediates more effective interactions while delivering them more efficiently—and doing so with little, if any, material variability in quality across the system. Metrics for an interface system's consistency are relevant from a front-office as well as a back-office perspective. From the standpoint of front-office operations, it's critical to ask: To what extent do customers perceive consistent levels of reliability and value delivered from one interface to the next? From the standpoint of back-office operations, it's important also to ask: To what extent do internal business systems support consistently reliable delivery of customer-perceived value from one interface to the

FIGURE 8-6

Interface System Scorecard: Efficiency Metrics

After completing your grids in tables 8-1 and 8-2, rate the performance of your company relative to its industry along the following metrics using a scale of 1 to 7, where 7 indicates that your company is a clear industry leader, 4 indicates that your company is on par with the industry, and 1 indicates that your company lags significantly behind the industry.

Efficiency Metrics	Score
1. Incremental cost per transaction (average of responses)	1 2 3 4 5 6 7
Retail store	1 2 3 4 5 6 7
Call center representative	1 2 3 4 5 6 7
VRU	1 2 3 4 5 6 7
Online site	1 2 3 4 5 6 7
Live e-mail	1 2 3 4 5 6 7
Automated e-mail (Extend list to include additional interfaces)	1 2 3 4 5 6 7
2. Average cost per interaction by usage occasion (average of responses)	1 2 3 4 5 6 7
List each usage occasion from Table 8-2	1 2 3 4 5 6 7
A low score for a usage occasion does not necessarily indicate poor efficiency if management has made an explicit choice to serve an occasion at higher cost to better meet customer needs. However, for any low score, consider whether it is due to explicit choice or inefficient distribution of interactions across available interfaces.	
3. Alignment of interfaces, customer segments, and usage occasions	1 2 3 4 5 6 7
Recall the figure from the previous chapter outlining what people and machines each do best. How well aligned are customer segments and usage occasions with interfaces reflecting what people and machines each do best? Has your company optimized this division of labor on an operating cost basis? For example, is your company using high-cost interfaces to mediate rote transactions? Are some customer segments concentrated users of higher-cost interfaces? Are there avenues to migrate customers to lower-cost interfaces?	

next? Use figure 8-7 to evaluate your interface system on consistency metrics.

Consistency from a customer perspective is in large measure a function of how effectively information is shared across interfaces in the system, particularly where your company aspires to customize interactions to individual customers or customer segments. A higher score on Question 2 indicates a robust capability in your organization to tailor interactions in consistently meaningful ways for customers across multiple interactions. A lower

FIGURE 8-7

Interface System Scorecard: Consistency Metrics

Rate the performance of your company relative to its industry along the following metrics using a scale of 1 to 7, where 7 indicates that your company is a clear industry leader, 4 indicates that your company is on par with the industry, and 1 indicates that your company lags significantly behind the industry.

Consistency Metrics	Score
1. Customer-perceived consistency of interactions (average of responses)	1 2 3 4 5 6 7
Extent to which:	
Information and representations are consistent across interfaces	1 2 3 4 5 6 7
Brand/corporate image is consistent across interfaces	1 2 3 4 5 6 7
Interactions reflect consistent quality over time	1 2 3 4 5 6 7
Interactions reflect consistent quality across interface types	1 2 3 4 5 6 7
Interactions reflect consistent quality across geographic locations	1 2 3 4 5 6 7
Contacts companywide are coherent and coordinated	1 2 3 4 5 6 7
2. Employee-perceived consistency in supporting interactions (average of responses)	1 2 3 4 5 6 7
Extent to which:	
Data is integrated across interfaces	1 2 3 4 5 6 7
Data is integrated across business units	1 2 3 4 5 6 7
Data is integrated across geographies	1 2 3 4 5 6 7
Data collected enables time-series comparison of metrics	1 2 3 4 5 6 7
Data is actively shared in real time with relevant business units	1 2 3 4 5 6 7
Service errors and service recovery are managed effectively	1 2 3 4 5 6 7
Business and IT systems are managed effectively	1 2 3 4 5 6 7

score on Question 1 is cause for concern, even with a higher score on Question 2, because customers' perceptions of consistency and reliability across your interface system will influence their decisions to do business with you in the future. If you score lower on any one of the sub-questions, you may wish immediately to pinpoint which interfaces in your system are the culprits. You may discover, for example, that your company has deployed just one poorly designed interface that is resulting in negative perceptions across your entire interface system. Not all problems your customers experience with your interface system will reflect systemic problems. In cases where consistency is sporadic at best, you may conclude that your company has failed on a strategic level either to define the objectives of the organization's interactions with its customers or to translate those objectives effectively into management actions and processes. Whatever the case, this exercise will help to identify problems and potential solutions.

Adaptability

An interface system must adapt to changing customer needs and expectations, competitor actions, and market contexts to sustain a company's competitive advantage. Adaptability is determined by organizational structures and processes. Under the best of circumstances, the front office serves as a listening post for the voice of the customer and market. Frontline service employees are best positioned to spot changes in customer desires or behaviors and convey critical information back to the organization; machines in frontline service positions—for example, those that monitor customer voices for changes in emotional tone or signs of distress—can do this, too. Such insights are invaluable to inform decisions regarding the testing and adoption of new interface technologies, and to ease the process of adoption by the organization and its customers. Use the scorecard in figure 8-8 to evaluate your interface system along adaptability metrics.

Achieving a high score on adaptability does not require that your company make the right decision about every technology it deploys and seamlessly integrate new interfaces into your interface system every single time. However, adaptability does require that your organization monitor customer and technology trends to determine which might result in opportunities to enhance your interface system. Your company will also need a process in place of managing its adaptation—including evaluation, testing, and deployment of new interfaces and other changes to the existing system. Without question, experimenting with technology is a risky prospect, and the ultimate risk profile your company embraces will depend on your customers' and employees' readiness to adopt new technologies and adapt to new ways of doing business. For

FIGURE 8-8

Interface System Scorecard: Adaptability Metrics

*Rate the performance of your company relative to its industry along the following
metrics using a scale of 1 to 7, where 7 indicates that your company is a clear
industry leader, 4 indicates that your company is on par with the industry, and
1 indicates that your company lags significantly behind the industry.*

Adaptability Metrics	Score
1. Tracking new customer, employee, and technology trends (average of responses)	1 2 3 4 5 6 7
Customer trends are actively followed	1 2 3 4 5 6 7
Employee trends are actively followed	1 2 3 4 5 6 7
Technology trends are actively followed	1 2 3 4 5 6 7
Feedback from customers and employees is appropriately processed	1 2 3 4 5 6 7
2. Responding to new customer, employee, and technology trends (average of responses)	1 2 3 4 5 6 7
Clear decision-making process or organization to manage new interface technology adoption	1 2 3 4 5 6 7
New interface technologies are actively incubated and developed	1 2 3 4 5 6 7
New interface technologies are tested effectively before rollout	1 2 3 4 5 6 7
Customers are appropriately educated and trained to use new interface technologies	1 2 3 4 5 6 7
Employees are appropriately educated and trained to use new interface technologies	1 2 3 4 5 6 7

every company, the right answer regarding degrees of innovation
in an interface system will be different.

Now that you have completed the scorecard exercise, determine
the average of your scores in each of the four sections—effectiveness, efficiency, consistency, and adaptability—and compute an
overall average score. You may summarize these scores in figure
8-9. While you must look at the individual scores to understand
more deeply the performance attributes of your entire interface
system, you can now draw certain conclusions based on your
overall score.

If your composite score is around 4, your interface system is at
parity with your competitors, but it does not yet qualify as a driver
of competitive advantage. Your company may have sophisticated

FIGURE 8-9

Interface System Scorecard: Summary

From each of the four sections of the scorecard, average your score for that section, then record your system's overall average score.

System Metrics	Average Score
1. Average **Effectiveness** Score (from figure 8-5)	_____
2. Average **Efficiency** Score (from figure 8-6)	_____
3. Average **Consistency** Score (from figure 8-7)	_____
4. Average **Adaptability** Score (from figure 8-8)	_____
Total Average Score	_____

competitors who manage interface systems particularly well, or your score may reflect performance shortfalls in one or more of the four key performance attributes. Ultimately, the goal of front-office reengineering is to deliver effectiveness and efficiency gains simultaneously, so it's critical to look for ways to improve individual interfaces or the consistency and adaptability of the entire system in ways that simultaneously raise interaction quality while lowering interaction costs. As we have seen, reconfiguring the front office through substitution of capital for labor is one way of achieving this goal; physical displacement of capital or labor is another. And hybrid interfaces, both people-led and machine-led, can generate unique sources of leverage and productivity for a company. As a rule, managers should first improve *execution* of interactions with customers within the existing interface system, then improve the overall interface system from the standpoint of *structural* changes.

If your company's composite score is below 4, your company's interface system is not operating at par with competitors. If your company has earned a particularly low score, you may wish to focus on improving the performance of individual interfaces—in both effectiveness and efficiency—before moving on to more complex systems issues. The scorecard results will provide a guide to determine where the problem areas lie, and then you may wish to set priorities according to the twin measures of urgency and impact.

If your average score is above 4, your company is outperforming the industry in management of its interface system—and in all likelihood it's achieving competitive advantage as a result. However, it's always useful to ask whether your company has proven itself smart or lucky. Are you performing better than the competition because of your company's superior interface system management or because of your competitors' poor management of their interface systems? If you are operating in an industry of weak interface system players, you may wish to evaluate your performance against a set of competitors in other industries known for superior relationship management. If you have scored particularly highly in system execution along dimensions of effectiveness and efficiency, you may wish to turn your attention to structural aspects (enabling consistency and adaptability) to ensure continuous improvement of your interface system as markets, customers, and technology evolve.

CONCLUSION

Beyond providing a set of scores, the elements of the scorecard should convey the key principles of sound management of interface systems. The scorecard should also point to areas of priority where appropriate data collection or processes may be lacking. Having now made this initial assessment of your interface system, you may proceed to the more complex challenges involved in specifying aspirations for customer interactions with your company and brand; aligning your interface system with organizational capabilities to realize this aspiration; articulating the interface system through improved interface and systems-level design; and activating your extended enterprise, including employees and customers, to maximize the efficiency, effectiveness, consistency, and adaptability of your interface system. These are the steps in the model we presented in chapter 7—the five A's of assessment, aspiration, alignment, articulation, and activation—that can optimize the interface system that will put a company's best face forward.

Notes

Chapter 1

1. David Wessel, "Productivity Gains: Never Bad, Even for American Workers," *New York Times*, 21 August 2003.

2. U.S. Department of Labor, Bureau of Labor Statistics, "National Cross-Industry Estimates of Employment and Mean Annual Wage for SOC Major Occupational Group," (Washington, DC: U.S. Department of Labor, Bureau of Labor Statistics, May 2003).

3. James A. Fitzsimmons and Mona J. Fitzsimmons, *Service Management: Operations, Strategy, and Information Technology* (New York: McGraw-Hill/Irwin, 2000), 5.

4. Organisation for Economic Co-operation and Development, *Main Economic Indicators: Basic Structural Statistics*, January 2004, <http://www.oecd.org/dataoecd/8/4/1874420.pdf?channelId=34247&homeChannelId=33715&fileTitle=Basic+Structural+Statistics> (accessed 22 March 2004).

5. To see employment breakdowns by position in each industry, see U.S. Bureau of Labor Statistics, "The 2004-05 Career Guide to Industries," 27 February 2004, <http://www.bls.gov/oco/cg/home.htm> (accessed 20 August 2004).

6. Texts of late that address customer experience management include: James H. Gilmore and B. Joseph Pine II, *The Experience Economy* (Boston: Harvard Business School Press, 1999); Bernd H. Schmitt, *Experiential Marketing* (New York: Free Press, 1999); Bernd H. Schmitt, *Customer Experience Management* (Hoboken, NJ: John Wiley & Sons, 2003); Colin Shaw and John Ivens, *Building Great Customer Experiences* (New York: Palgrave Macmillan, 2002); Shaun Smith and Joe Wheeler, *Managing the Customer Experience*, 1st ed. (New York: Financial Times Prentice Hall, 2002).

7. For example, Gilmore and Pine write, "We believe buyers will purchase transformations according to the set of eternal principles the seller seeks to embrace—what together they believe will last. . . . According to our own worldview, there can be no [next] economic offering because perfecting people falls under the province of God" (*The Experience Economy*, 206).

8. Quotations are found, respectively, in Gilmore and Pine, *The Experience Economy*, chapter 2; Schmitt, *Experiential Marketing*, chapters 3 through 7; and Smith and Wheeler, *Managing the Customer Experience*, chapter 2.

9. For example, see Louis Uchitelle, "Good Economy. Bad Job Market. Huh?" *New York Times*, 14 September 2003.

10. Fred Brock, "Who'll Sit at the Boomers' Desks?" *New York Times*, 12 October 2003.

11. Shaw and Ivens define this kind of emotional intelligence as "self-awareness, self-regulation, empathy, motivation, and social skills," *Building Great Customer Experiences*, 105.

12. John Hood, "Blessings of Liberty: The Market Approach to Job Training," *Policy Review*, no. 77 (May–June 1996).

13. "Alexandrians made the first accurate measure of the earth's circumference in the third century B.C. Also, in roughly the same period, an astronomer named Aristarchus concluded that the earth revolved around the sun (this was 1,800 years before Copernicus) and two anatomists concluded that 'the brain was the center of the nervous system and the seat of intelligence.'" Finally, "the engineer Heron of Alexandria, in his work 'Pneumatica,' laid out the principles of steam power." Braudel as quoted in Alexander Stille, "Resurrecting Alexandria," *New Yorker*, 8 May 2000, 90–99.

14. Bernard Lewis, *What Went Wrong: Western Impact and Middle Eastern Response* (Oxford, England: Oxford University Press, 2001).

15. American Customer Satisfaction Index, "American Customer Satisfaction Index, 1994 to Q4 2003," ACSI Web site, <http://www.theacsi.org/national_scores.htm> (accessed 22 March 2004).

16. Technology Quarterly, "The Gentle Rise of the Machines," *The Economist*, 13–19 March 2004, 29–30.

17. Ibid.

18. For example, Delta invested more than $1.5 billion to get its digital nervous system in place. Deborah Gage and John McCormick, "Delta's Last Stand," *Baseline Magazine*, 1 April 2003, <http://www.baselinemag.com/article2/0,3959,1016969,00.asp?> (accessed 22 March 2004).

19. For further background on the role of organizational context on the success of IT investment, see Erik Brynjolfsson, Lorin M. Hitt, and Shinkyu Yang, "Intangible Assets: Computers and Organizational Capital," working paper 138, Center for eBusiness at MIT, Cambridge, MA, October 2002; and Erik Brynjolfsson, "The IT Productivity Gap," *Optimize*, no. 21 (July 2003).

20. Benjamin Farmer and Robin Goad, "Voice Automation: Past, Present and Future," Intervoice and Datamonitor white paper, July 2003.

21. Charles Haddad, "The Web Smart 50: Krispy Kreme," *BusinessWeek*, 24 November 2003, 88.

22. Cliff Edwards, "The Web Smart 50: Charles Schwab," *BusinessWeek*, 24 November 2003, 92.

23. Roger O. Crockett, "The Web Smart 50: Progressive Insurance," *BusinessWeek*, 24 November 2003, 98.

24. Jon E. Hilsenrath, "Behind Surging Productivity: The Service Sector Delivers," *Wall Street Journal*, 7 November 2003.

25. Jim Kerstetter, "The Web Smart 50: Sutter," *BusinessWeek*, 24 November 2003, 100.

26. Hilsenrath, "Behind Surging Productivity."

27. For example, see Craig Karmin, "'Offshoring' Can Generate Jobs in the U.S.," *Wall Street Journal*, 16 March 2004.

28. Jack E. Triplett and Barry Bosworth, "Productivity Measurement Issues in Services Industries: 'Baumol's Disease' Has Been Cured," *Federal Reserve Bank of New York Economic Policy Review*, September 2003, 23–33.

29. Hilsenrath, "Behind Surging Productivity."

30. John L. King, "IT Responsible for Most Productivity Gains," *Computing Research News* 15, no. 4 (2003): 1, 6.

31. Dale W. Jorgenson, Mun S. Ho, and Kevin J. Stiroh, "Lessons from the U.S. Growth Resurgence," paper prepared for the First International Conference on the Economic and Social Implications of Information Technology, held at the U.S. Department of Commerce, Washington, DC, 27–28 January 2003 and updated PowerPoint materials of the same title, November 2003; Dale W. Jorgenson, "Information Technology and the G7 Economies," *World Economics*, 4, no. 4, October–December 2003, 139–169; Dale W. Jorgenson, "Economic Growth in the Information Age," PowerPoint presentation to Universita Bocconi, 5 December 2003; Erik Brynjolfsson and Lorin M. Hitt, "Computing Productivity: Firm-Level Evidence," working paper 4210-01, MIT Sloan/working paper 139, eBusiness at MIT, June 2003.

32. Paul Lukas, "On Wings of Commerce," *Fortune*, 8 March 2004.

33. Erik Brynjolfsson, "The IT Productivity Gap," *Optimize* 21 (July 2003). For a more scholarly treatment of this topic, see Timothy F. Bresnahan, Erik Brynjolfsson, and Lorin M. Hitt, "Information Technology, Workplace Organization, and the Demand for Skilled Labor: Firm-Level Evidence," *Quarterly Journal of Economics* 117 (2002): 339–76.

34. For example, see Jorgenson, "Information Technology and the G7 Economies"; Brynjolfsson and Hitt, "Computing Productivity"; Bresnahan, Brynjolfsson, and Hitt, "Information Technology."

35. Norbert Weiner, *The Human Use of Human Beings* (Boston: Houghton Mifflin, 1954).

36. For an example of such a scenario, see Bill Joy, "Why the Future Doesn't Need Us," *Wired*, April 2000.

Chapter 2

1. For an example, see Jeremy Rifkin, *The End of Work*, paperback ed. (New York: G. P. Putnam and Sons, 1995).

2. See Southwest Airlines, "We Weren't Just Airborne Yesterday," Southwest Airlines Web site, <http://www.southwest.com/about_swa/airborne.html> (accessed 22 March 2004); Continental Airlines News Release, "Continental Airlines Launches International E-Ticket and Announces Interline E-Ticket Development," Continental Airlines Web site, 14 May 1998, <http://www.continental.com/vendors/default.asp?SID=935D3CF9FF5E49A3826046E9DCDB64C3&s=&i=%2Fcompany%2Fnews%2F1998%2D05%2D14%2D01%2Easp> (accessed 22 March 2004).

3. Louis Kraar, "Your Next PC Could Be Made in Taiwan," *Fortune*, 8 August 1994.

4. Alfred D. Chandler with the assistance of Takashi Hikino, *Scale and Scope: The Dynamics of Industrial Capitalism* (Cambridge, MA: Belknap/Harvard, 1990); Alfred D. Chandler, *The Visible Hand: The Managerial Revolution in American Business* (Cambridge, MA: Harvard University Press, 1980).

5. Pallavi Gogoi, "Financial Services—I Love You—but I'm Leaving You," *BusinessWeek*, 21 July 2003, <http://www.businessweek.com/@@dtx93ocQhDos@g8A/magazine/content/03_29/c3842016_mz003.htm> (accessed 30 March 2004).

6. James H. Gilmore and B. Joseph Pine II, *The Experience Economy* (Boston: Harvard Business School Press, 1999); Michael J. Wolf, *The Entertainment Economy: How Mega-Media Forces Are Transforming Our Lives* (New York: Crown Business, 1999); Don Tapscott, *The Digital Economy: Promise and Peril in the Age of Networked Intelligence* (New York: McGraw-Hill, 1996); Shoshana Zuboff and James Maxmin, *The Support Economy: Why Corporations Are Failing Individuals and the Next Episode of Capitalism* (New York: Viking Penguin, 2002).

7. Kerry Capell and Gerry Khermouch, "Hip H&M: The Swedish Retailer Is Reinventing the Business of Affordable Fashion," *BusinessWeek*, 11 November 2002.

8. See, "Chrysler Aims to Lead in Quality, Productivity," *Autotech Daily*, 6 October 2003, <http://www.autotechdaily.com/pdfs/T10-06~1.PDF> (accessed 30 March 2004); National Automobile Dealers Association (NADA), "Annual NADA DATA Report: Auto Retail Industry Maintains Strength in 2003," NADA Online 11 May 2004, <http://www.nada.org/Content/NavigationMenu/Newsroom/News_Releases/2004/ind_05_11_04.htm> (accessed 11 August 2004).

9. Tom Osenton, *The Death of Demand: Finding Growth in a Saturated Global Economy* (Upper Saddle River, NJ: Financial Times Prentice Hall, 2004).

10. Ad Age, "100 Leading National Advertisers," AdAge.com, 28 June 2004, <http://www.adage.com/news.cms?newsId=40827.htm> (accessed 8 August 2004).

11. Ibid.

12. Ed Garsten, "Auto Incentive War Heats Up," *Detroit News*, 6 August 2004 <http://www.detnews.com/2004/autoinsider/0408/09/c01-234506.htm> (accessed 11 August 2004).

13. Jacqueline Doherty, "Turning on the Lights," *Barron's*, 24 February 2003, <http://online.wsj.com/barrons/article/0,SB104586995434454583,00.html?mod=b_this_weeks_magazine_main> (accessed 2 December 2003).

14. Spencer E. Ante, "The New Blue," *BusinessWeek*, 17 March 2003, <http://www.businessweek.com/magazine/content/03_11/b3824001_mz001.htm> (accessed 22 March 2004).

15. Technology Quarterly, "The Gentle Rise of the Machines," *The Economist*, 13–19 March 2004, 30.

16. Ibid.

17. NTT DoCoMo, "Subscriber Growth," NTT DoCoMo Web site, <http://www.nttdocomo.com/companyinfo/subscriber.html> (accessed 11 August 2004).

18. "InStat/MDR Forecasts 118 Million 3G Subscribers in China by 2008," in *3G Network Deployments in China, TelecomDirect News*, June 2004.

19. Antone Gonsalves, "RFID Warning for Wal-Mart Suppliers," *Information Week* Web site, 21 November 2003, <http://www.informationweek.com/story/showArticle.jhtml?articleID=16400307> (accessed 2 December 2003).

20. James Pearce, "New Insulator Strengthens Future of Moore's Law," ZDNet Australia, 5 January 2004, <http://news.zdnet.co.uk/hardware/chips/0,39020354,39118854,00.htm> (accessed 22 March 2004); Gordon Moore, "No Exponential Is Forever . . . but We Can Delay 'Forever,'" presentation at International Solid State Circuits Conference, 10 February 2003.

21. International Federation of Robotics and the United Nations, *World Robotics 2003* (Geneva: United Nations, 2003), 2.

22. Ibid., 286.

23. Ibid.

24. Pyo Jae-yong, "Make Way for the Machines," *Joong Ang Daily*, 13 August 2003, <http://joongangdaily.joins.com/200308/13/200308130141385079900091009101.html> (accessed 2 December 2003).

25. Ibid.

26. Siemens AG and Hefter Cleantech, "Make Way! Robots Are Revolutionizing Cleaning," Siemens AG Web site, 25 May 2000, <http://w4.siemens.de/en2/html/press/press_release_archive/releases/2000052501e.html> (accessed 12 December 2003).

27. Youngme Moon, "Sony AIBO: The World's First Entertainment Robot," Case 9-502-010 (Boston: Harvard Business School, 2003).

28. Stephan Wilkinson, "uDrive Me Crazy," *Popular Science*, <http://www.popsci.com/popsci/auto/article/0,12543,386094,00.html> (accessed 22 March 2004).

29. Moon, "Sony AIBO."

30. Louise Knapp, "A Way out of Automated Phone Hell," Wired News, 10 February 2004, <http://www.wired.com/news/gizmos/0,1452,62184,00.html> (accessed 22 March 2004).

31. Nikkei Electronics, "Honda Develops a Humanoid Robot That Can Walk More Naturally," Nikkei Electronics Online, 21 November 2000, <http://ne.nikkeibp.co.jp/english/2000/11/1120asimo_d-cehtml> (accessed 12 December 2003).

32. Sony, "Sony Develops Small Biped Entertainment Robot," Sony Web site, 19 March 2002, <http://news.sel.sony.com/pressrelease/2360> (accessed 12 December 2003).

33. This is the kind of alchemical reaction that science fiction has long explored; in the mass-market form, it's the subject—with a great deal of poetic license—of Steven Spielberg's 2001 film, *AI: Artificial Intelligence*.

34. Honda, "Event Report. Vol. 1: ASIMO Rings Opening Bell at NYSE," Honda Worldwide Web site, 15 February 2002, <http://world.honda.com/ASIMO/event/wreport_01.html> (accessed 12 December 2003).

35. BBC, "Robot Attends Czech State Dinner," BBC News Web site, 21 August 2003, <http://news.bbc.co.uk/1/hi/world/asia-pacific/3170061.stm> (accessed 30 April 2004).

36. Yudhijit Bhattacharjee, "Getting to Know All About You: Making Robots More Like Us," *New York Times*, 6 March 2003.

37. Richard Shim, "Palm V Skips into BlackBerry Patch," CNET News.com, 17 October 2001, <http://news.com.com/2100-1040-274574.html? legacy=cnet> (accessed 12 December 2003).

38. Mary Meeker and Chris DePuy, *The Internet Report* (Morgan Stanley, 1996).

39. Ben Macklin, "Choice of Broadband Tech + Access Provider = Broadband Growth," *eMarketer*, 3 April 2003.

40. David Carnoy, "Q&A with Sega: The Game Developer's Take on Xbox Live vs. PS2 Online," CNET Reviews, 14 November 2002, <http://electronics. cnet.com/electronics/0-3622-8-20665788-2.html> (accessed 12 December 2003); Entertainment Software Association, "Essential Facts About the Computer and Video Game Industry," Entertainment Software Association Web site, May 2004, <http://www.theesa.com/EFBrochure.pdf> (accessed 30 March 2004). (The Entertainment Software Association was formerly the Interactive Digital Software Association.)

41. RocSearch Ltd., *Video Game Market* (London: RocSearch Ltd., November 2003); Chuck Kahn, "Worldwide Box Office," Worldwideboxoffice.com, 2003, <http://www.worldwideboxoffice.com> (accessed 22 March 2004).

42. eMarketer, "Broadband Worldwide 2004: Subscriber Update Report," *eMarketer Report*, 4 April 2004.

43. Cliff Edwards et al., "Special Report: Digital Homes," *BusinessWeek*, 21 July 2003, <http://www.businessweek.com/@@Squ7VYcQkqcJnQsA/magazine/ content/03_29/b3842088.htm> (accessed 22 March 2004).

44. United Nations, *World Robotics* 2003, 285–320.

Chapter 3

1. James R. Beninger, *The Control Revolution: Technological and Economic Origins of the Information Society* (Cambridge, MA: Harvard University Press, 1986); Shoshana Zuboff, *In the Age of the Smart Machine: The Future of Work and Power*, reprint ed. (New York: Basic Books, 1988); Alfred D. Chandler, *The Visible Hand: The Managerial Revolution in American Business* (Cambridge, MA: Belknap Press, 1980).

2. Thanks to them, machine breaking in Britain became a capital offense.

3. David Bacon, "Unions Fear 'War on Terror' Will Overcome Right to Strike," InterPress Service, listed on Common Dreams News Center, 10 August 2002, <http://www.commondreams.org/headlines02/0810-02.htm> (accessed 30 March 2004).

4. David E. Sanger and Steven Greenhouse, "President Invokes Taft-Hartley Act to Open 29 Ports," *New York Times*, 9 October 2002.

5. Pete Engardio, Aaron Bernstein, and Manjeet Kripalani, "The New Global Job Shift," *BusinessWeek*, 3 February 2003.

6. Figures from John C. McCarthy, analyst for Forrester Research Inc., Cambridge, MA, cited in Engardio, Bernstein, and Kripalani, "The New Global Job Shift."

7. Ibid.

8. For example, see National Association of Chain Drug Stores (NACDS), "Remarks of Mary Sammons, President and CEO of Rite Aid Corporation at

the NACDS Pharmacy & Technology Conference" in Philadelphia, PA, 25 August 2003, <http://www.nacds.org/wmspage.cfm?parm1=3213> (accessed 29 December 2003).

9. Pete Bowles, "Union Lists Imperiled Token Booths," *New York Newsday*, 22 April 2003; Pete Donohue, "Token Booth Closing Time: Machines to Replace the Clerks at 177 Sites," *New York Daily News*, 14 January 2003; Luis Perez, "MTA Proposes Metrocard Hikes," *New York Newsday*, 29 July 2004.

10. James L. Heskett, "Lessons in the Service Sector," *Harvard Business Review*, March–April 1987.

11. Richard Gibson, "Machine Takes Orders in McDonald's Test," *Wall Street Journal*, 11 August 1999.

12. Borders, "Borders Group Investor Presentation," Borders Web site, June 2004 <http://phx.corporate-ir.net/phoenix.zhtml?c=65380&p=irol-presentations> (accessed 13 August 2004).

13. Kiosk.com, "Web-Based Kiosks + Free Delivery to Stores = Surging Sales," Kiosk.com, 2 May 2003, <http://www.kioskcom.com/articles_detail.php?ident=1735> (accessed 30 March 2004).

14. Wal-Mart has self-checkout lanes in most of its three thousand stores, and Home Depot and Kmart, as well as grocers H-E-B and Cub, are deploying similar checkout technology in their markets. Bill Brewer, "Welcome to the Do-It-Yourself Revolution," Scripps Howard News Service, 20 March 2000.

15. Joseph Agnese, "Industry Surveys: Supermarkets and Drugstores," *Standard & Poor's*, 19 December 2002, 3.

16. NACDS, "Remarks of Mary Sammons."

17. Roger W. Ferguson, Jr. and William L. Wascher, "Distinguished Lecture on Economics in Government: Lessons from Past Productivity Booms," *Journal of Economic Perspectives* 18, no. 2 (spring 2004): 3–28.

18. James Parks, "The Future of Manufacturing and America's Middle Class," (Washington, DC: American Federation of Labor and Congress of Industrial Organization, April 2004).

19. James C. Cooper and Kathleen Madigan, "A Temporary Reprieve for Manufacturing," *BusinessWeek*, 29 September 2003, 33–34.

20. Michael Hammer, "Reengineering Work: Don't Automate, Obliterate," *Harvard Business Review*, July–August 1990.

21. For an academic review of the productivity paradox literature from 1996, see Erik Brynjolfsson, "The 'Productivity Paradox': A Clash of Expectations and Statistics," *Communications of the ACM* [Association for Computing Machinery], December 1993.

22. John A. Quelch, "Chemical Bank: The Pronto System," Case 9-584-089 (Boston: Harvard Business School, 1984, revised 1988).

23. James Heskett, W. Earl Sasser, Jr., Leonard Schlesinger, "Achieving Breakthrough Service" video series, (Boston: Harvard Business School Video Series, 1992).

24. Doug Andersen, "Financial Services Delivery Channels," Tower-Group, February 2002.

25. Ross Anderson, "Perspectives—Automatic Teller Machines," Cambridge University class notes posted on newsgroup alt.security, 8 December 1992.

26. Large regional banking corporation's proprietary research.

27. R. B. Breen and M. Zimmerman, "Rapid Onset of Pathological Gambling in Machine Gamblers," *Journal of Gambling Studies* 18 (2002): 1. Bob Breen, the Director of the Rhode Island Gambling Treatment Program, found that playing video slots is the most addictive form of gambling in history, taking only one year to hook players, versus close to four years for other types of gambling.

28. Joe Ashbrook Nickell, "Welcome to Harrah's," *Business 2.0*, April 2002, <https://www.business2.com/subscribers/articles/mag/print/0,1643,38619,00.html> (accessed 30 March 2004).

29. Harrah's, "Statistical Data as of December 31, 2003," Harrah's Web site, 31 December 2003, <http://www.harrahs.com/about_us/announcements_news/HET_STATS.pdf> (accessed 30 March 2004).

30. Roy Furchgott, "Wi-Fi Technology Moves from Storeroom to Store," *New York Times*, 25 August 2003.

31. Robert Goldfield, "Wells Fargo Branches to Offer More Than Banking," *Business Journal of Portland*, 24 November 1997, <http://portland.bizjournals.com/portland/stories/1997/11/24/newscolumn4.html> (accessed 30 March 2004).

32. E*Trade, "E*TRADE Opens Flagship Super-Store in New York," E*Trade Web site, 4 April 2001, <https://us.etrade.com/e/t/home/PressStory?ID=STORYID%3Dpr040401_PR&year=2001&month=04> (accessed 30 March 2004).

33. ING Direct, "About Us: Cafés," ING Direct Web site, <http://home.ingdirect.com/about/about.html> (accessed 30 March 2004).

34. Soumitra Dutta and Sameer Oundhakar, "ING Direct: Redefining Direct Banking," Case 302-184-1 (Fontainebleau, France: INSEAD, 2002).

35. Theodore Levitt, "Marketing Myopia," *Harvard Business Review*, September–October 1975.

36. James L. Heskett et al., "Putting the Service-Profit Chain to Work," *Harvard Business Review*, March–April 1994; James L. Heskett, W. Earl Sasser, Jr., and Leonard A. Schlesinger, *The Value Profit Chain: Treat Employees Like Customers and Customers Like Employees* (New York: Free Press, 2002).

37. Frederick F. Reichheld, *The Loyalty Effect* (Boston: Harvard Business School Press, 1996).

38. James L. Heskett, W. Earl Sasser, Jr., Leonard A. Schlesinger, *The Service Profit Chain* (New York: Free Press, 1997).

39. Frederick F. Reichheld and W. Earl Sasser, Jr., *Zero Defections: Quality Comes to Services*, HBR OnPoint enhanced ed. (Boston: Harvard Business School Publishing, 2000); Reichheld, *The Loyalty Effect*.

40. Michael E. Porter, *Competitive Advantage: Creating and Sustaining Superior Performance* (New York: Free Press, 1998).

41. Valerie Zeithaml, A. Parasuraman, and Leonard Berry, *Delivering Quality Service: Balancing Customer Perceptions and Expectations* (New York: Free Press, 1990).

42. The term originates from the computer industry standard called the Open Systems Interconnection (OSI) model, which was proposed to ensure that users could communicate with software systems in accessible formats using intuitive commands. OSI is a seven-layer model for describing communication

between two points on a network, defined by the International Standards Organization.

43. Donald A. Norman, "Emotion & Attractive," *Interactions*, July–August 2002; Donald A. Norman, *Emotional Design: Why We Love (or Hate) Everyday Things* (New York: Basic Books, 2003); F. Gregory Ashby, Alice M. Isen, and U. Turken, "A Neuropsychological Theory of Positive Affect and Its Influence on Cognition," *Psychological Review* 106, no. 3 (1999): 529–50; Alice M. Isen, "An Influence of Positive Affect on Decision Making in Complex Situations: Theoretical Issues with Practical Implications," *Journal of Consumer Psychology* 11, no. 2 (2001): 75–85.

44. Virginia Postrel, *Substance of Style* (New York: HarperCollins, 2003), 196.

45. Siegfried Giedion, *Mechanization Takes Command* (New York: W. W. Norton & Company, 1969).

Chapter 4

1. Nation's Restaurant News, "Panera, In-N-Out Burger Score Top Marks in Survey," Entrepreneur.com, 15 March 2004, <http://www.entrepreneur.com/Your_Business/YB_SegArticle/0,4621,314693-1——,00.html> (accessed 29 March 2004); as we discuss later, In-N-Out has recently started deploying employees with PDAs to speed order entry.

2. Segment on Washington Mutual and Bank One, *Today Show*, 1 September 2003.

3. Jennifer Merritt, "Improv at the Interview," *BusinessWeek*, 3 February 2003.

4. Steven Greenhouse, "Going for the Look but Risking Discrimination," *New York Times*, 13 July 2003.

5. Bill Birchard, "Hire Great People Fast," *Fast Company*, August–September 1997, 132.

6. Edmond Rostand, *Cyrano De Bergerac*, reissue ed. (New York: Bantam, 1950).

7. Robert L. Simons and Hilary A. Weston, "Nordstrom: Dissension in the Ranks? (A)," Case 9-191-002 (Boston: Harvard Business School, 1990); Robert L. Simons, "Nordstrom: Dissension in the Ranks? (B)," Case 9-192-027 (Boston: Harvard Business School, 1991).

8. Andrew Harper, "The World's Best Hotels, Resorts, and Hideaways: Our 21st Annual Survey of Sophisticated Travelers," *Andrew Harper's Hideaway Report*, September 2002. The Four Seasons chain placed fourteen properties on the global top forty, with Ritz-Carlton garnering only three. The study, according to *Forbes*, is based on polling of "2,500 sophisticated travelers, more than 85% of whom have the title president, CEO, owner or partner." Alexandra Kirkman, "Hotel for All Seasons," *Forbes Global Magazine*, 28 October 2002.

9. Kirkman, "Hotel for All Seasons."

10. Quotation from Four Seasons Annual Report, 2002.

11. Marriot Web site, Corporate Information, <http://marriott.com/corporateinfo/default.mi> (accessed 14 August 2004).

12. Kirkman, "Hotel for All Seasons."

13. Alexandra Kirkman, "Find and Teach the Best People," *Forbes Global Magazine*, 28 October 2002.

14. W. Earl Sasser, Jr., Thomas O. Jones, and Norman Klein, "Ritz-Carlton: Using Information Systems to Better Serve the Customer," Case 9-395-064 (Boston: Harvard Business School, 1994, revised 1999).

15. Southwest's first self-serve terminals were called LUV Machines and were introduced in 1979. Southwest Airlines, "We Weren't Just Airborne Yesterday," Southwest Airlines Web site, <http://www.iflyswa.com/about_swa/airborne.html> (accessed 30 March 2004). Its online bookings now make up 49 percent of its tickets, as opposed to 5 percent for American (Susan Carey, "Jet-Blue, One of Few U.S. Airlines to Buck the Downturn, Files for $125 Million IPO," *Wall Street Journal*, 13 February 2002). For a discussion of Southwest's strategic advantage, see Michael E. Porter, "What Is Strategy?" *Harvard Business Review*, November–December 1996.

16. Market capitalization data as of 16 August 2004. Yahoo! Finance, "Industry-Transportation-Airline-Company list," Yahoo! Finance Web site, <http://biz.yahoo.com/p/airlinconameu.html> (accessed 16 August 2004).

17. JetBlue has recently departed from this strategy, with an order of one hundred Embraer 190 regional jets, which will assuredly complicate its operations and raise costs to some extent.

18. Comment by Herb Kelleher to Jim Wooten, CBS news correspondent, in "Herb and His Airline," *60 Minutes*, 19 April 1997.

19. Brian Rogers and Nirmalya Kumar, "EasyJet: The Web's Favorite Airline," Case GM 873 (Lausanne, Switzerland: International Institute for Management Development, 2000).

20. Bureau of Transportation Statistics (BTS) "BTS Releases First Quarter 2004 Airline Financial Data; Regional Passenger Airlines Report Highest Rate of Domestic Profit," 14 June 2004, <http://www.bts.gov/press_releases/2003/bts029_03/html/bts029_03.html> (accessed 13 August 2004).

21. Frances X. Frei, "Rapid Rewards at Southwest Airlines," Case 9-602-065 (Boston: Harvard Business School, 2003).

22. Ibid. In 2004, Southwest underwent labor strife under James Parker, the CEO who replaced Kelleher. Kelleher ultimately had to step in to reach a resolution with flight attendants, and Parker resigned. This underscores the challenges of relying on a people-dominant interface system, particularly when it is dependent on a specific person like Kelleher. For more information, see Aude Lagorce, "Southwest CEO Lands in the Hotseat," *Forbes*, 29 March 2004; "Southwest Airlines CEL Resigns," *Air Wise News*, 15 July 2004.

23. JetBlue reviewer, "The Jetblue Experience in the Blackout of '03," epinions.com, 23 August 2003, <http://www.epinions.com/content_11002718 1700> (accessed 29 March 2004); Andrew Compart et al., "In Blackout, Points of Light," *Travel Weekly*, 25 August 2003, <http://www.findarticles.com/cf_dls/m3266/34_62/108115111/p1/article.jhtml> (accessed 29 March 2004); Whitney Tilson, "JetBlue's Challenges," *The Motley Fool*, Fool.com Web site, 19 September 2003, <http://www.fool.com/news/commentary/2003/commentary 030919wt.htm> (accessed 29 March 2004).

24. Airsafe.com, "Fatal Events and Fatal Event Rates by Airline Since 1970," 13 January 2003, <http://www.airsafe.com/airline.htm> (accessed 29 March 2004).

25. Tilson, "JetBlue's Challenges."

26. Russell Grantham, "Delta's Song All Tuned Up for Battle," *Atlanta Journal Constitution*, 9 November 2003.

27. Anthony J. Rucci, Steven P. Kirn, and Richard T. Quinn, "The Employee-Customer-Profit Chain at Sears," *Harvard Business Review*, January–February 1998.

28. Home Depot, "About the Home Depot," Home Depot Web site, <http://www.homedepot.com/HDUS/EN_US/corporate/about/about.shtml> (accessed 30 March 2004).

29. Dan Maher and Dan O'Brien, "The Journey from Good to Great: Office Depot (C)," OU-081C (New York: Omnicom University and the Omnicom Group, Inc, 2003).

30. In August 2004, Toys "R" Us announced it might leave the toy business altogether. Constance L. Hays, "Toys "R" Us Says It May Leave the Toy Business," *New York Times*, 12 August 2004.

31. For a discussion of employees' intellectual skills, see John Hood, "Blessings of Liberty: The Market Approach to Job Training," *Policy Review*, no. 77 (May–June 1996).

Chapter 5

1. Statistic about the volume of business from Hertz, "Hertz History," Hertz corporate Web site, <http://www.hertz.com/about_05/index.html> (accessed 23 December 2003).

2. Hertz, "Hertz Firsts: 1995," Hertz corporate Web site, <http://www.hertz.com/about_05/index.html> (accessed 23 December 2003).

3. As reported, "Hilton, Hyatt, and Sheraton were among the chains that tried and abandoned express check-in kiosks [in 1999], but Hilton is betting that better technology and travelers' increased familiarity will make them more popular with customers this time." Barbara DeLollis, "Hilton to Set Up Self-Serve Kiosks in New York, Chicago," *E-Commerce Times*, 4 September 2003.

4. Kortney Stringer, "How to Have a Pleasant Trip: Eliminate All Human Contact," *Wall Street Journal*, 31 October 2002.

5. The first discussion of white-collar and blue-collar machines of which we're aware appeared in an examination of machine technology in industrial settings in Shoshana Zuboff, *The Age of the Smart Machine* (New York: Basic Books, 1989).

6. Joseph Agnese, "Industry Surveys: Supermarkets and Drugstores," *Standard & Poor's*, 19 December 2002, 3.

7. Hans P. Moravec, "Robots, After All," *Communications of the ACM*, October 2003, 90–97.

8. In one experiment, a classical music audience was asked to listen intently to three pieces of music, one written by a music professor, Eric Larson, in the style of J. S. Bach and another written by a computer programmed with the same aim. The audience concluded, erroneously, that Larson's piece had been written by the computer, and the computer's piece was decreed that of Bach. Kurzweil's Cybernetic Poet program can also compose haiku after "reading" poems by John Keats and Wendy Dennis. Ray Kurzweil, *The Age of Spiritual Machines* (New York: Penguin Group, 1999), 160, 163.

9. IBM, "Deep Blue Wins Match," IBM Research Web site, 11 May 1997, <http://www.research.ibm.com/deepblue/home/may11/story_1.html> (accessed 30 March 2004).

10. Ray Kurzweil, "Deep Fritz Draws: Are Humans Getting Smarter, or Are Computers Getting Stupider?" KurtweilAI.net, 19 October 2002, <http://www.kurzweilai.net/articles/art0527.html> (accessed 30 March 2004).

11. Hans P. Moravec, "Encyclopedia Britannica Article," Hans P. Moravec Web site, July 2003, <http://www.frc.ri.cmu.edu/~hpm/project.archive/robot.papers/2003/robotics.eb.2003.html> (accessed 30 March 2004); Moravec, "Robots, After All."

12. See the figure which lists human strengths versus machine strengths in Erik Brynjolfsson, "The IT Productivity Gap," *Optimize* no. 21 (July 2003), <http://www.optimizemag.com/showArticle.jhtml?articleID=17700941> (accessed 30 March 2004).

13. John Tierney, "Shop Till Eggs, Diapers, Toothpaste Drop," *New York Times*, 28 August 2002.

14. Michael Singer, "E-books to Zoom into New Airport E-stores," *Internet News*, 25 May 2001, <http://siliconvalley.internet.com/news/article.php/773841> (accessed 30 March 2004).

15. Joe McKendrick, "Exchange Takes the Lead—This Year," *ENT News*, 14 October 2002, <http://www.entmag.com/news/article.asp?EditorialsID=5540> (accessed 30 March 2004).

16. Michelle Krebs, "What's the Word on the T-Bird?" *New York Times*, 24 June 2001.

17. Ad Age, "100 Leading National Advertisers," AdAge.com, 23 June 2003, <http://www.adage.com/images/random/lna03.pdf> (accessed 22 March 2004).

18. For example, "some analysts predict that one of the Big Three will disappear in the next decade under mounting pressure from the likes of Toyota and Honda." Danny Hakim, "A Family's 100-Year Car Trip," *New York Times*, 15 June 2003.

19. "My Mother the Car" briefly aired on NBC from September 1965 to September 1966.

20. Associated Press, "Google Surges 18% in NASDAQ Debut," *MSNBC News*, 19 August 2004 <http://www.msnbc.msn.com/id/5743246/> (accessed 19 August 2004).

21. David Pogue, "Meeting the Googlers," *New York Times*, 25 March 2004.

22. Jef Raskin, *The Humane Interface*, 1st ed. (Upper Saddle River, NJ: Addison-Wesley, 2000).

23. Marvin Minsky, *The Society of Mind* (New York: Simon & Schuster, 1988), 163. Also, Rosalind Picard of MIT wrote, "The latest scientific findings indicate that *emotions play an essential role in rational decision making, perception, learning, and a variety of other cognitive functions*." She observed that, as early as 1967, Herbert Simon identified what Daniel Goleman called "emotional intelligence" as essential to cognition, indicating that "a general theory of thinking and problem solving must incorporate the influences of emotion." Rosalind Picard, *Affective Computing* (Cambridge, MA: MIT Press, 1997), x, 1–3.

24. Martyn Williams, "Sony Shows High End, High Price Gadgets," *PC World*, 10 June 2003.

25. Stuart Elliott, "Music Is at the Center of New Toyota Campaign," *New York Times*, 28 August 2001.

26. Jeffrey Zaslow, "Oh No! My TiVo Thinks I'm Gay," *Wall Street Journal*, 4 December 2002.

27. Ken Belson and Brian Bremner, *Hello Kitty: The Remarkable Story of Sanrio and the Billion Dollar Feline Phenomenon* (New York: John Wiley & Sons, 2003).

28. "Tellme Tells All," *Next Innovator*, 25 November 2002, <http://www.technologyreports.net/nextinterface/?articleID=924> (accessed 30 March 2004).

29. This phone call reached the magazine processing center in Salt Lake City, Utah, on 11 May 2003.

30. Richard S. Wallace, "From Eliza to A.L.I.C.E.," A.L.I.C.E. AI Foundation Web site, <http://www.alicebot.org/articles/wallace/eliza.html> (accessed 30 March 2004).

31. Byron Reeves and Clifford Nass, *The Media Equation* (Cambridge, England: Cambridge University Press, 1996).

32. The UCLA Center for Communication Policy, "The UCLA Internet Report—Surveying the Digital Future," February 2003, <http://ccp.ucla.edu/pdf/UCLA-Internet-Report-Year-Three.pdf> (accessed 30 March 2004).

33. Jon Van, "Machine-to-Machine Talk Not Stuff of Fiction; More Affordable Commercial Uses," *Chicago Tribune*, 2 September 2003.

34. As of April 2004, Orange had discontinued Wildfire service while developing other voice recognition services.

35. Kurzweil, *The Age of Spiritual Machines*, 20–25.

36. Rodney A. Brooks, *Flesh and Machines: How Robots Will Change Us* (New York: Pantheon Books, 2002).

37. Hans P. Moravec, *Mind Children: The Future of Robot and Human Intelligence* (Cambridge, MA: Harvard University Press, 1994).

38. Picard, *Affective Computing*.

39. Ibid., 35.

40. Ibid., 32.

41. Jeremy Rifkin, excerpt from the introduction of *The End of Work: Five Years Later*, paperback ed. (New York: Penguin, 2000), <http://www.jobsletter.org.nz/art/rifkin05.htm> (accessed 29 March 2004). Rifkin observes: "The cheapest workers in the world—from the factory floor to the professional suites—will not be as cheap and efficient as the intelligent software and wetware coming on line to replace them. By the mid-decades of the 21st century, computers, robotics, biotechnologies, and nano-technologies will be able to produce cheap and abundant basic goods and services for the world's human population, employing a fraction of the world's human labour in the process. . . . In the year 2050, less than five percent of the human population on earth—working with and alongside intelligent technology—will be required to produce [everything] needed by the human race."

42. Ibid.

43. Joanna Glaner, "How Robots Will Steal Your Job," *Wired*, 5 August 2003.

44. Marshall Brain, "Robotic Nation," marshallbrain.com, <http://marshallbrain.com/robotic-nation.htm> (accessed 30 March 2004).

Chapter 6

1. See Stewart Brand, *The Media Lab: Inventing the Future at MIT*, reprint ed. (New York: Penguin USA, 1988).

2. Steve Mann, "An historical account of the 'WearComp' and 'WearCam' inventions developed for applications in 'Personal Imaging,'" Published in IEEE Proceedings of the first ISWC, Cambridge, Massachusetts, 13–14 October 1997, 66–73, <http://wearcam.org/historical/index.html> (accessed 3 February 2004).

3. Nextel (featuring BlackBerry), "Doubtful and Definite" advertisement, *BusinessWeek*, 2 June 2003, 13.

4. Lorraine B. Diehl and Marianne Hardart, *The Automat: The History, Recipes, and Allure of Horn & Hardart's Masterpiece* (New York: Clarkson Potter Publishers, 2002).

5. John Carey et al., "Point, Click . . . Fire," *BusinessWeek*, 7 April 2003; Stan Crock et al., "The Doctrine of Digital War," *BusinessWeek*, 7 April 2003.

6. Wal-Mart Annual Report, 2003.

7. Jim Hopkins, "Wal-Mart's Influence Grows," *USA Today*, 29 January 2003.

8. James L. Heskett, W. Earl Sasser, Jr., and Christopher W. L. Hart, *Service Breakthroughs* (New York: Free Press, 1990).

9. Stephen Rose, "Going for the Look, but Risking Discrimination," *New York Times*, 13 July 2003.

10. When flying first class recently, we saw a flight attendant ask a passenger intemperately, "Have you been *beveraged* yet?" Why was she so pushy, especially with a first-class passenger? She snapped at us, "Most of them don't deserve good service—they upgraded from coach with miles."

11. Christopher W. L. Hart, James L. Heskett, and W. Earl Sasser, Jr., "The Profitable Art of Service Recovery," *Harvard Business Review*, July–August 1990.

12. W. Earl Sasser, Jr., Thomas O. Jones, and Norman Klein, "Ritz-Carlton: Using Information Systems to Better Serve the Customer," Case 9-395-064 (Boston: Harvard Business School, 1999).

13. Wyndham International, a middle-market hotel chain, has achieved similar success by implementing such a system. Its ByRequest guest-recognition program has substituted more effective personalization of service for reward points. See Gabrielle Piccoli and Lynda M. Applegate, "Wyndham International: Fostering High-Touch with High-Tech," Case 9-803-092 (Boston: Harvard Business School, 2002).

14. James L. Heskett and Kenneth Ray, "Fairfield Inn (A)," Case 9-689-092 (Boston: Harvard Business School, 1989); James L. Heskett and Kenneth Ray, "Fairfield Inn (B)," Case 9-692-005 (Boston: Harvard Business School, 1993).

15. Sandelman & Associates, "Fast-Food Awards of Excellence for 2003," Sandelman Web site, 1 March 2004, <http://www.sandelman.com/news/pdf/FFAward.pdf> (accessed 19 August 2004).

16. Richard Meryhew, "Making Fast Food That Much Faster," *Star Tribune*, 27 July 2004.

17. Dina Berta, "Survey: Link Between Economy, Lower Turnover Rates," *Nation's Restaurant News*, 29 March 2004, 14.

18. For more information on the company, see <http://www.liveperson.com> (accessed 23 August 2004).

19. Jeffrey F. Rayport, Mary Connor, and Thomas A. Gerace, "Weather Services Corp.," Case 9-396-052 (Boston: Harvard Business School, 1995, revised 1998).

20. Jim Rutenberg and Micheline Maynard, "TV News That Looks Local, Even if It's Not," *New York Times*, 2 June 2003.

21. "The typical age range of First Direct customers was between 25 and 54, with the average being 42. Customers' average earnings were 1.5 times the national average, and they were twice as financially active as the average person in the United Kingdom. They are income rich and time poor, but not necessarily asset rich. They are technophiles, with twice the national propensity to take up new technology." "Telephone Banking Pioneer First Direct Embraces the Internet," *Bank Marketing International*, 28 August 2003.

22. That said, First Direct aimed to derive maximum leverage from the infrastructure of its corporate parent. According to its current CEO, Alan Hughes, "We use HSBC's infrastructure as much as possible, for example its ATMs, cheque-clearing, its back-office IT systems, and its membership of organisations such as the U.K. debit card scheme. But the customer-facing side of First Direct uses our own systems." "Telephone Banking Pioneer First Direct."

23. Jean-Marie Dru, *Disruption: Overturning Conventions and Shaking Up the Marketplace*, 1st ed. (New York: John Wiley & Sons, 1996); Jean-Marie Dru, *Beyond Disruption: Changing the Rules in the Marketplace*, 1st ed. (New York: John Wiley & Sons, 2002).

24. Jeffrey F. Rayport and Dickson L. Louie, "First Direct (A)," 9-897-079 (Boston: Harvard Business School, 1997, revised 1998).

25. First Direct, "Three in a Row—First Direct Is Best for Customer Service Again," First Direct Web site, 27 May 2004, <http://www.firstdirect.com/press/releases/release90.htm> (accessed 19 August 2004).

26. "Telephone Banking Pioneer First Direct."

27. Linda Robertson, "Taking Account of Experience: Internet Bank Is Looking for Mature Workers to Use Their Know-How to Help Customers," *Evening Times* (Glasgow), 10 March 2003.

28. Each call center "has a subsidised on-site nursery. There are also chill-out zones, courses such as creative writing and languages and a café. Flexible benefits (including health insurance, cinema tickets, and discounted gym membership) are worth between £875 and £2,186 a year, and there is a profit-related bonus. Rewards range from boxes of chocolates to an entire day of treats when First Direct won an award as the 'most recommended bank in the world.'" Cited in "100 Best Companies to Work for 2003," *Sunday Times* (Britain), 2 March 2003.

29. "Telephone Banking Pioneer First Direct."

30. For example, classic type A personalities had no interest in talking about sports or the weather before slowly getting to the tasks at hand; rather, they wanted to transact and get off the phone as quickly as humanly possible. More relaxed account holders, such as university students and retirees, derived enormous satisfaction from their conversations, regardless of whether they did anything more than check their balances for the day. Customers in a bad humor, especially one that was generated by a previous interaction with the

bank, required different interpersonal approaches than customers who were fully satisfied with their recent interactions. (This was First Direct's approach to managing service recovery.)

31. See Rayport and Louie, "First Direct."

32. "Generating Customer Satisfaction the First Direct Way," *Bank Marketing International*, 3 September 2002.

33. "First Direct Employs Virtual Sales Agent," *Bank Marketing International*, 26 February 2003, 4.

34. "Telephone Banking Pioneer First Direct."

35. Ibid.

36. Rayport and Louie, "First Direct (A)"; Jeffrey F. Rayport and Carrie L. Ardito, "First Direct (B)," Case 9-898-145 (Boston: Harvard Business School, 1998).

***Chapter* 7**

1. William Grimes, "When the Cashier Is You," *New York Times*, 7 April 2004.

2. Michelle Higgins, "Grocery Shopping Enters a New Age," *Wall Street Journal*, 30 March 2004.

3. Robert Sofman, senior vice president for e-business at Charles Schwab & Co., as quoted in Bob Tedeschi, "E-Commerce Report," *New York Times*, 19 April 2004.

4. Schwab brochure, "Welcome to Schwab Independent Investing Signature," Charles Schwab & Co., Inc., 2004.

5. Advertisement, "American Express Financial Advisors," *BusinessWeek*, 22 March 2004, opposite page 74.

6. Analyst Report, "Liberty Media," Morgan Stanley, 7 January 2004; Press release, "IAC Reports Q4 2003 Results," IAC/InterActiveCorp, 9 February 2004; "Annual Report," Liberty Media 2003.

7. Ibid. Figure based on U.S. revenue/U.S. active purchasing customers; Dara Khosrowshahi and Tom McInerney, IAC InterActiveCorp presentation, Bear Stearns 17th Annual Media, Entertainment and Information Conference, 9 May 2004, <http://customer.nvglb.com/BEAR002/030804a_cy/pdf/interactive.pdf> (accessed 2 April 2004); interviews with QVC executives by author, August 2004.

8. Interviews with QVC executives by author, August 2004.

9. Jeffrey F. Rayport and Elizabeth B. Glass, "TV-Home Shopping Wars: QVC and Its Competitors," Case number 9-395-014 (Boston: Harvard Business School, 1994, revised 1995).

10. "TV Home Shopping Players Project Optimistic Outlook for Next Year," *The Electronic Times* (Korea), 24 December 2003.

11. Jeffrey F. Rayport and Dickson L. Louie, "QVC, Inc.," Case number 9-897-050 (Boston: Harvard Business School, 1996, revised 1997).

12. Ibid.

13. QVC Web site, "Business Overview," <http://www.qvc.com/asp/frameset.asp?dd=nav/navhqwel.html&nest=/mainhqwel.html?tmp=hp&cont=sn> (accessed 2 April 2004).

14. Alexandra Kaptik, "Strategies for Securing a Home-Shopping Spot," Startup Journal Case Study from *Wall Street Journal* online, 1 April 2003, <http:www.startupjournal.com/columnist/casestudy/20030401-casestudy.html> (accessed 2 April 2004).

15. Interviews with QVC executives by author, August 2004.

16. John Hunter, "I Want My QVC," *CEO Magazine*, 1 June 2003 <http://www.cio.com/archive/060103/perspectives.html> (accessed 1 April 2004); Rama Ramaswami, "Out of Service," *Operations & Fulfillment*, 1 January 2003, <http://opsandfulfillment.com/ar/fulfillment_service_2/> (accessed 1 April 2004); Sasha Issenberg, "Getting Ready for Prime Time," *Inc. Magazine*, November 2003, <http://www.inc.com/magazine/20031101/spotlight.html> (accessed 1 April 2004).

17. Nancy Jeffries, "Alternative Channels: The Global Village Is Driving the Shape of Shopping by Catering to the Female Imperative," *Global Cosmetic Industry* no. 8, vol. 171 (2003): 42; Vickie Chachere, "HSN Marks 25 Years on TV," *Associated Press Online*, 17 July 2002.

18. Interviews with QVC executives by author, tape recording, West Chester, PA, 5 September 2002 and 13–14 September 2002.

19. Ibid.

20. Interviews with QVC executives by author, August 2004.

21. John M. Higgins, "Top 25 TV Networks," *Broadcasting and Cable*, 1 December 2003, <http://www.broadcastingcable.com/index.asp?layout=article&articleid=CA338554&display=Features&doc_id=130498> (accessed 1 April 2004, subscription required).

22. In the early days of television, the UVF spectrum assignments went to first movers—NBC, ABC, and CBS—via their affiliates. Independent TV stations in local markets got double-digit channels in the UHF spectrum. Direct-broadcast satellite and digital television eliminate the relevance of such assignments.

23. Interviews with QVC executives by author, August 2004.

24. Interviews with QVC executives by author, tape recording, West Chester, PA, 13–14 November 2002.

25. An exception to the taking title rule occurs with the Web business; the site features a million SKUs, of which 60,000 have been featured on air, and QVC does not take title to those additional items. Instead, it utilizes drop-ship arrangements. However, logistics executives have shown that they can control third-party fulfillment sufficiently to maintain the low error rates, using software from CommerceHub.

26. Hunter, "I Want My QVC."

27. Ibid.

28. QVC must also seek favorable channel "adjacencies" by placing itself on cable dials in "neighborhoods" within viewers' so-called clicker rotations. That means negotiating channel assignments contiguous to the broadcast networks' affiliated stations. Popular channels are in the low numbers, and so QVC must be among them to capture channel surfers.

29. Interviews with QVC executives.

30. Interviews with QVC executives; Sandra Dolbow, "Rose Goes Outside, Inside for QVC Brands," *Brandweek*, 31 July 2000.

31. Revenue, EBITDA, and employee figures from: Analyst Report, "Liberty Media," Morgan Stanley, 7 January 2004; Press Release, "Liberty Media Corporation Provides Fourth Quarter and Full Year Supplemental Financial Information and 2004 Guidance," Liberty Media, 15 March 2004, <http://www.libertymedia.com/press_release/default.htm> (accessed 1 April 2004); Sears, "Annual Report," 2003; Press Release, "Sears Reports Fourth Quarter 2003 Results," Sears, 29 January 2004, <http://phx.corporate-ir.net/phoenix.zhtml?c=63737&p=irol-newsArticle&ID=489542&highlight> (accessed 1 April 2004); Wal-Mart, "Annual Report" 2003; Analyst Report, "Wal-Mart Stores Inc."; McDonald Investments Inc., 2 March 2004.

32. Liberty Media Press Release, "Liberty Media Corporation to Acquire QVC, Inc.," 3 July 2003.

33. QVC is a registered service mark of ER Marks, Inc. The views, personal impressions, and opinions expressed in this book are those of the authors and not necessarily those of the management of QVC, Inc.

34. Liberty Media, "Annual Report," 2003; InterActiveCorp, "Annual Report," 2003.

35. QVC Web site, "Corporate Facts"; InterActiveCorp, "Annual Report."

36. Mark Albright, "HSN Takes Calls to the Philippines, *St. Petersburg Times*, 19 November 2003.

37. InterActiveCorp, "Annual Report"; Liberty Media, "Annual Report; interviews with QVC executives."

38. Home Shopping Network Web Site, "Company Information," <http://www.hsn.com/corp/info/default.aspx> (accessed 1 April 2004).

39. Khosrowshahi and McInerney, IAC InterActiveCorp presentation; Issenberg, "Getting Ready for Prime Time."

40. Figures based on U.S. revenue and customers.

41. Liberty Media, "Annual Report"; Press Release, "IAC Reports Q4 2003 Results," InterActiveCorp, 9 February 2004.

42. A number of threads on HSN's bulletin boards discuss this. For example, customers remarked that host Colleen Lopez and guest Adrienne (of signature Club A cosmetics) talked over each other, apparently ignoring one another.

43. As of April 2004.

44. Broadcast viewing observations on QVC and HSN show variety, March 2004.

45. Albright, "HSN Takes Calls to the Philippines."

46. Ibid.; Ramin Ganeshram, "Excellence: Outsourcing—Making the Right Call," *CRM Magazine*, February 2003, <http://www.destinationcrm.com/articles/default.asp?ArticleID=2853> (accessed 1 April 2004).

47. Hunter, "I Want My QVC."

48. David Myron, "Delivering on Its Promise—CRM Is Turning Call Centers into Profit Centers," *CRM Magazine*, April 2003, <http://www.destinationcrm.com/articles/default.asp?ArticleID=2992> (accessed 1 April 2004).

49. Home Shopping Network has since improved these figures over the past two years; Khosrowshahi and McInerney, IAC InterActiveCorp presentation.

50. A series of HSN bulletin board postings detail these experiences. One passionate posting is titled "Ultrex-Customer Experience." It describes cus-

tomer frustration over the many products in the line that malfunction within a few months. As one of the largest retailers of Ultrex, HSN does not accept returns after thirty days—or help customers communicate with the manufacturer.

Chapter 8

1. For example, see Frederick F. Reichheld, "The One Number You Need to Grow," *Harvard Business Review*, December 2003.

Visualization of Key Terms

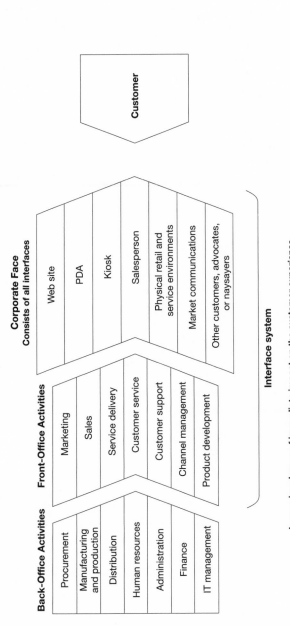

Back-Office Activities

Procurement

Manufacturing
and production

Distribution

Human resources

Administration

Finance

IT management

Front-Office Activities

Marketing

Sales

Service delivery

Customer service

Customer support

Channel management

Product development

Corporate Face
Consists of all interfaces

Web site

PDA

Kiosk

Salesperson

Physical retail and
service environments

Market communications

Other customers, advocates,
or naysayers

Customer

Interface system

Increasing degree of immediate impact on the customer experience

GLOSSARY OF KEY TERMS

While historically managers have thought about activities within a company according to a value-chain model, which illustrates the ways a company creates value as it moves from raw materials or factor inputs through sales and marketing activities to customers, we focus on those activities with greatest leverage on company-customer interaction, customer relationship management, and ultimately customer experience. As a result, we complete the value chain with an additional set of front-office elements (see figure G-1).

Back office: The primary locus within a company of internally focused activities, processes, and systems that support the creation and distribution of core offerings, including (1) the manufacturing and production of products and services and (2) supporting functions such as administration, human resources, finance, procurement, and distribution.

Corporate face: The elements of interface systems that represent the visible points of connection between a company and its customers and that impinge directly on a company's management of interactions and relationships with customers.

Customer experience: Customers' perception and interpretation of their interactions and relationships with a company, its product or services, and its brand, as mediated by its interfaces or interface systems.

Front office: The primary locus within a company of externally focused activities, processes, and systems that support the positioning of customer offerings, delivery of services, and the management of customer interactions and relationships, including (1) marketing, sales, and customer service; (2) supporting functions pertaining to go-to-market channels and technologies; and (3) customer-facing interfaces.

Interface: Any entity—whether human or machine, product or media—that companies deploy externally or internally to manage interactions or relationships with customers, markets, and employees. Interfaces are entities that serve to connect a company to its customers or employees, and include such entities that might be controlled by third parties. Interfaces range from physical environments a company may create for its customers (retail stores, service counters) to elements inside those environments (salespeople, interactive kiosks, point-of-sale promotion) to elements belonging to third-party influencers (Web sites, word-of-mouth referrals).

Interface system: The unique system of interfaces and related activities and processes that a company's front office actively and strategically deploys as a means of managing customer interactions and relationships in optimally efficient and effective ways.

Index

Page numbers ending in *f* or *t* indicate a figure or table, respectively.

ACSI (American Customer Satisfaction Index), 8, 33
activation of interface systems, 177, 203–204
adaptability of interface system, 215, 216f, 225–226. *See also* scorecard for interface system
advertising spend, 35, 124
affective appeal
  as a driver of relationship management, 43–47, 89–90 (*see also* four drivers for technology as relationship manager)
  emotional attributes of a service interface, 129–135
  potential for complex bonds to form, 47
  socially perceptive reasoning by machines, 45
  users' development of emotional attachment, 44, 45–46
Affective Tigger, 44–45
The Age of Spiritual Machines, 118
AIBO robotic dog, 41, 42–44, 137
airline industry, low-cost
  differentiated interface system based on people, 104–106
  economic success of low-cost airlines, 101–102
  focus on simplicity of operations, 102–103
  interface benefits from reduced complexity of operations, 106–107
  use of a machine interface, 29
alignment of front-office capabilities, 177, 201–202
Amazon.com, 12, 126
American Customer Satisfaction Index (ACSI), 8, 33
Ananova, 132
anchor interface, 89
Apple Computer, 114
articulation of a plan of execution, 177, 202–203
ASIMO by Honda, 45, 46
aspiration for customer interactions, 177, 199–201
assessment of customers' current experiences, 177, 195–199
ATMs, 68, 70–71, 74–75
audits. *See* interface audit
automobile industry, 123–125. *See also* BMW; GM; Toyota
  advertising spend, 35, 124
  complexity of systems, 43
  connectivity and, 51
  dynamics of push and pull interfaces, 123–124
  labor protests against, 56
  margin compression, 35
  overcapacity, 35

back office, defined, 10, 249
banking industry. *See also* ATMs;
    Citibank; ING Direct
    hybrid-interface productivity
        model example (*see* First
        Direct)
    hybrid interfaces adoption, 77–78
Banryu, 41
Barrett, Colleen, 104
Baumol, William, 17
Baumol's Disease , 17
BlackBerry, 147
BMW, 43
Borders, 60
Bowersox, Bob, 184
Brain, Marshall, 141
brands, 114–115, 124
Braundel, Fernand, 7
Briggs, Doug, 180
broadband and connectivity, 48
Brooks, Rodney, 140
Brynjolfsson, Erik, 17

call centers
    off-shoring jobs and, 57
    QVC versus HSN, 188, 197
    recruitment and training at First
        Direct, 166
central-casting, 161
Charles Schwab & Co., 13, 174–175
chess played by machines, 118
China, 39
Citibank, 70
Clear Channel Communications, 161
Club Quarters, 116
cognition, 89, 90
cognitive attributes of a service
    interface, 125–129
commoditization
    challenges in orchestrating optimal
        interactions, 32–33
    interface imperative, 32
    interfaces as an opportunity to
        create value, 31–32
    three-six-one world, 31–32
competitive advantage
    customer satisfaction index, 8

establishing and sustaining, 11
evaluating status (*see* interface
    audit; scorecard for interface
    system)
front-office interfaces and (*see*
    front-office reengineering)
industrial revolution and, 2–3
interactions and relationships
    defined, 5, 6f
interfaces as the basis of, 1–2
labor scarcity and the industrial
    revolution, 7–8
labor scarcity paradox, 5–7
machines used to manage customer
    relationships, 9
managers' neglect of systems as
    sources of, 176
new focus on customer experience,
    4–5
services' role in businesses, 2
summary, 19–21
connectivity
    convergence of devices and
        networks, 51–52
    as a driver of relationship
        management, 48–52, 89, 90 (*see
        also* four drivers for technology
        as relationship manager)
    forms, 48–49
    localized in small businesses, 92–94
    penetration into residences, 49–51
    spread of, 48
consistency of interface system, 215,
    216f, 222–225. *See also* scorecard
    for interface system
customer experience
    defined, 249
    relationship with interface systems,
    4–5
customer relationship management
    (CRM), 58–59
corporate face, defined, 249

Davenport, Thomas, 64
Deep Blue, 118
Deep Fritz, 119
Delta, 107

device proliferation, 37–39, 40f. *See also* four drivers for technology as relationship manager

Diller, Barry, 180

domestic service robots, 9, 41, 131, 137. *See also* robotics

easyJet, 101, 103

economics and the service sector
  capital-for-labor substitutions protests, 55–56
  current service sector employment, 3
  drivers of reengineering revolution, 11–12, 13f
  economics of interface systems, 70–71
  economic success of low-cost airlines, 101–102
  impact of front-office reengineering, 13–15
  implications of growth in machine-based jobs, 141–142
  labor scarcity and the industrial revolution, 7–8
  labor scarcity paradox, 5–7
  manufacturing and services productivity parallels, 62–63
  shift from agricultural to industrial employment, 2–3

effectiveness of interface system, 1, 215, 216f, 217–218, 219f. *See also* scorecard for interface system

efficiency of an interface system, 1, 215, 216f, 218–222. *See also* scorecard for interface system

Eliza, 133

emotional attributes and machines
  emotional appeal of human or animal forms in media, 132
  research examining human responses to interactive technologies, 133–134
  role of emotion in interactions, 130–131, 140–142
  service quality's influence on purchase decisions, 134–135
  users' development of emotional attachment, 44, 45–46

Emotional Design, 84

e-ticketing, 29

exchange businesses, 127

Fairfield Inn, 153–155

Fanuc, 41

fashion industry, 34

fast-food industry, 60, 87–88, 155–158

FedEx, 14

First Direct
  attainment of personal service feeling, 168–169
  company background, 162–163
  customer information tracking levels, 166–167
  employee recruitment and training, 165–166
  focus on new customer segment through advertising, 163–165
  impact of combining customer data, 167–168
  interface drivers leveraged, 170
  machines' contribution to productivity and customer base, 169–170

Five A's model, 176–178. *See also* activation of interface systems; alignment of front-office capabilities; articulation of a plan of execution; aspiration for customer interactions; assessment of customers' current experiences

four drivers for technology as relationship manager, 36–52. *See also* affective appeal; connectivity; device proliferation; intelligence and interactivity

Four Seasons Hotels and Resorts, 96–97, 98–99, 100

front-office capabilities
  applied to QVC and HSN (*see* QVC versus HSN)

front-office capabilities *(continued)*
  process description, 194
  stages of interface system design,
    177
  summary, 205
front office, defined, 10, 249
front-office reengineering
  advantages to interactions free of
    human assistance, 12
  in banking industry, 77–78
  company examples of front-office
    machines use, 13–14
  described, 57–58
  economic drivers of, 11–12, 13f
  economic impact of, 13–15
  emergence and embracing of,
    64–66
  gaming industry example, 73–77
  human impact, 15–16
  hybrid interfaces, 69
  interface systems overview, 66–67,
    68f
  IT investment payback question,
    17–19
  machine-dominant interfaces,
    68–69
  optimization challenge in, 69–70
  people-dominant interfaces, 67–68
  pros and cons of reengineering, 72
  service sector productivity gains
    from technology, 16–17
  strategic thinking about interface
    systems, 70–72
front-office revolution
  capital-for-labor substitutions
    protests, 55–56
  challenges of *(see* service
    management challenges)
  customer relationship
    management, 58–59
  front and back office defined, 10
  guideposts for *(see* front-office
    revolution guiding principles)
  implications, 59
  machines as service providers,
    83–84
  off-shoring jobs and, 57
  productivity gains, 59–64

reengineering *(see* front-office
    reengineering)
  summary, 85–86
front-office revolution guiding
    principles
  better jobs for people, 22–23
  division of labor changes, 21–22
  goal of simplifying systems, 25
  interface systems as basis of
    competition, 23–24
  interface systems as brands, 24
  machines trained to act like people,
    22
  relationship between employees
    and customers, 25–26
  seeing technology as neutral, 23

gaming industry. *See* Harrah's
    Entertainment
  machine-dominant interface, 73–74
  multiple-interface system
    advantages, 74–77
  people-dominated interface, 73
Gap, 34
GE Medical Systems, 35–36
General Motors (GM), 51, 123–124
  advertising spend, 35, 124
  sit-down strike, 56
Google, 128

Hammer, Michael, 64
Harrah's Entertainment, 74–77
Hawkins, Jeff, 47
Hennes & Mauritz, 34
Hertz, 115–116
Hilton Hotels, 116
Home Shopping Network. *See* QVC
    versus HSN
home-shopping sector. *See* QVC;
    QVC versus HSN
Horn & Hardart Automat, 148
hospitality industry, 96–101. *See also*
    Fairfield Inn; Four Seasons
    Hotels and Resorts; Hilton
    Hotels; Hotel Okura; Hyatt
    Hotels; Ritz-Carlton Hotels

consistent quality through hiring
and training, 98–99
differentiation strategies based on
service, 96–99
focus on interactions and
relationships, 94–96
maintaining individual personality
and creativity, 100–101
results of programmed behavior, 100
technology used to differentiate
how staff serves guest, 150–153
technology used to shape
employees' attitudes and
behaviors, 153–155
Hotel Okura, 98
HSN. *See* QVC versus HSN
human capital, 18
human-machine interactions
cognitive attributes of a service
interface, 125–129
emotional attributes of a service
interface, 129–135
flowchart of interactions, 120f
integrating the attributes, 136–139
physical attributes as a service
interface, 121–125
synaptic attributes of a service
interface, 135–136
Hyatt Hotels, 117
hybrid interface systems. *See* people
combined with machines

IBM Global Services, 36
Idei, Nobuyuki, 130
Immelt, Jeffrey, 36
industrial revolution, 2–3, 7–8
ING Direct, 77–78, 169
In-N-Out Burger, 87–88
interactions defined, 5, 6f
intelligence and interactivity, 40–43
interface, xvii–xviii, 5. *See also* Charles
Schwab, First Direct, QVC
archetypes, 68f
defined, 249
interface advantage, 30, 36–37, 176
interface audit
evaluation questions, 212–215

purpose of assessment, 208–209
system inventory steps, 209–211
system scorecard (*see* scorecard for
interface system)
interface imperative, 32
interface system, xviii, 251. *See also*
scorecard for interface system
adaptability of 215, 216f, 225–226
defined, 251
interface technology
assessment approach (*see* interface
audit)
capability as the basis of
competitive advantage, 37
collaboration between humans and
machines example, 29
commoditization and (*see*
commoditization)
hybrid interfaces (*see* people
combined with machines)
interface imperative, 52–53
machines' legacy of substitution
for humans, 28–29
managing systems (*see* managing
interface systems)
margin compression, 35–36
product life-cycle acceleration,
33–34
services as basis of revenues, 36
summary, 52–53
superior interface systems
examples (*see* First Direct;
QVC)
International Longshore and
Warehouse Union, 56
Isen, Alice, 84
IT investment payback question,
17–19

Japan, 39
JetBlue Airways, 101, 102, 104
Jorgenson, Dale, 17

Kasparov, Garry, 118
Kawasaki, 41
Kismet, 45

Kramnik, Vladimir, 118–119
Krispy Kreme, 13
Kroger, 61
Kurzweil, Ray, 118

labor
    compensation index, 12
    outsourcing/off-shoring jobs, 21,
        51, 57, 58
    scarcity, 5–8 (*see also*
        outsourcing/off-shoring jobs)
    strikes, 56–57
Lands' End, 109
Levine, Kathy, 183–185
LG Home Shopping, 178–179
Luddites, 55

machine-dominant interfaces
    business implications of interactive
        machines, 140
    compensating for employee's lack of
        expertise or knowledge, 113–114
    described, 68–69
    dynamics of human-machine
        interactions (*see* human-machine
        interactions)
    economic implications of growth
        in machine-based jobs, 141–142
    future trends, 139–140
    human versus machine
        intelligence, 118–119
    potential of white-collar machines,
        117–118
    products as the faces and voices of
        brands, 114–115, 124
    role of emotion in interactions,
        140–142
    scope of tasks now done by
        machines, 115–117
    summary, 142–144
machines
    chess playing and, 118
    collaboration with humans
        example, 29
    company examples of front-office
        machines use, 13–14

contribution to humanizing the
    face of business, 15–16
contribution to productivity and
    customer base at First Direct,
    169–170
as the dominant interface (*see*
    machine-dominant interfaces)
emotional attributes and (*see*
    emotional attributes and
    machines)
in gaming industry (*see* gaming
    industry)
human response to a positive
    interface, 84
hybrid interfaces (*see* people
    combined with machines)
legacy of substitution for humans,
    28–29
machine-to-machine
    communication, 135–136
as service providers, 83–84
socially perceptive reasoning by, 45
task replacement of humans
    examples, 92
trained to act like people, 22
used to manage customer
    relationships, 9
machines supported by people
    human personality amplified by
        machines, 161–162
    leverage model, 159–161
    productivity model example (*see*
        First Direct)
    sources of value created, 158–159
managing interface systems
    affective appeal, 43–47
    device proliferation, 37–39, 40f
    intelligence and interactivity,
        40–43
    managers' neglect of systems as
        sources of competitive
        advantage, 176
    stages of system design (*see* Five A's
        model)
    summary, 204–205
    superior interface systems examples
        (*see* First Direct; QVC)
    supermarket systems, 173–174

consistent quality through hiring
and training, 98–99
differentiation strategies based on
service, 96–99
focus on interactions and
relationships, 94–96
maintaining individual personality
and creativity, 100–101
results of programmed behavior, 100
technology used to differentiate
how staff serves guest, 150–153
technology used to shape
employees' attitudes and
behaviors, 153–155
Hotel Okura, 98
HSN. *See* QVC versus HSN
human capital, 18
human-machine interactions
cognitive attributes of a service
interface, 125–129
emotional attributes of a service
interface, 129–135
flowchart of interactions, 120f
integrating the attributes, 136–139
physical attributes as a service
interface, 121–125
synaptic attributes of a service
interface, 135–136
Hyatt Hotels, 117
hybrid interface systems. *See* people
combined with machines

IBM Global Services, 36
Idei, Nobuyuki, 130
Immelt, Jeffrey, 36
industrial revolution, 2–3, 7–8
ING Direct, 77–78, 169
In-N-Out Burger, 87–88
interactions defined, 5, 6f
intelligence and interactivity, 40–43
interface, xvii–xviii, 5. *See also* Charles
Schwab, First Direct, QVC
archetypes, 68f
defined, 249
interface advantage, 30, 36–37, 176
interface audit
evaluation questions, 212–215

purpose of assessment, 208–209
system inventory steps, 209–211
system scorecard (*see* scorecard for
interface system)
interface imperative, 32
interface system, xviii, 251. *See also*
scorecard for interface system
adaptability of 215, 216f, 225–226
defined, 251
interface technology
assessment approach (*see* interface
audit)
capability as the basis of
competitive advantage, 37
collaboration between humans and
machines example, 29
commoditization and (*see*
commoditization)
hybrid interfaces (*see* people
combined with machines)
interface imperative, 52–53
machines' legacy of substitution
for humans, 28–29
managing systems (*see* managing
interface systems)
margin compression, 35–36
product life-cycle acceleration,
33–34
services as basis of revenues, 36
summary, 52–53
superior interface systems
examples (*see* First Direct;
QVC)
International Longshore and
Warehouse Union, 56
Isen, Alice, 84
IT investment payback question,
17–19

Japan, 39
JetBlue Airways, 101, 102, 104
Jorgenson, Dale, 17

Kasparov, Garry, 118
Kawasaki, 41
Kismet, 45

Published by Basic Books,
A Member of the Perseus Books Group

Designed by Lisa Kreinbrink
Set in 10-point Meridien

Library of Congress Cataloging-in-Publication Data

Large, David Clay.
And the world closed its doors : the story of one family
abandoned to the Holocaust / by David Clay Large.
    p.   cm.
Includes bibliographical references and index.
ISBN 0-465-03808-5
1. Schohl, Max, 1884-1943.   2. Jews—Germany—Flörsheim—
Biography.   3. Holocaust, Jewish (1939-1945)—Germany—
Flörsheim.   4. Schohl family.   5. Schohl, Max, 1884-1943—
Correspondence.   6. Flörsheim (Germany)—Biography.   I. Title.

DS135.G5S35755   2003
940.53'18'0922—dc21

2002153720

03 04 05 / 10 9 8 7 6 5 4 3 2 1

To
Käthe Schohl Wells
and in memory of
Dr. Max Schohl

man Jews were applying for visas under a national quota system that remained highly restrictive. Nonetheless, both Max and Julius were confident that the Schohls' visa applications would be successful. After all, Max was a talented chemist and proven entrepreneur; his wife Liesel and his two daughters, eighteen-year-old Helene and fifteen-year-old Käthe, were all willing and anxious to work hard in their prospective new home. Were not Max and his family just the kind of new citizens that America needed?

After roughly two years of effort—of increasingly desperate attempts to find a way through or over the "paper walls" of American immigration policy—Max had to accept the grim reality that America did not, in fact, really need or want the Schohls. In 1939 the American consulate at Stuttgart allotted Max and his family a number within the annual German immigration quota (part of a racially based national quota system for all immigrants to the United States) that would not have allowed them to leave Germany for America until 1944, a delay that Max understandably believed would be fatal.

Despairing of reaching America, Max next sought to emigrate to England, where he had business contacts who promised to help him gain admission to that nation. But here too he ran into government-mandated impediments, which became insurmountable with the outbreak of war in September 1939. Facing the Gestapo-imposed deadline to leave Germany or face arrest, the Schohl family chose Yugoslavia as a temporary refuge in hopes of using that Balkan state as a springboard to emigration to Brazil. But before they could secure entry permits, Brazil too closed its doors. With the Nazi conquest of the Balkans in spring 1941, the Schohls' Yugoslavian refuge turned into a trap. A year after Max's deportation his wife and daughters were ordered back to Germany to work as slave laborers. They survived this ordeal, and by emigrating to America shortly after the war Käthe Schohl was finally able to make the passage that was denied to her father.

As the foregoing outline makes evident, the Schohl story constitutes a small part of a much larger story, or collection of stories. One of the bigger pictures here clearly involves the reaction of deeply assimilated German Jews to the gradually escalating Nazi racial persecution in their country. In putting off his bid to abandon Germany to such a late date, Max was allowing his love of country to obscure his survival instincts, a miscalculation that he shared with thousands of his co-religionists. Like all too many of them, he believed that the German people were too sophisticated, too cultured, to tolerate for long the barbarous brownshirted gangsters who had taken over his country in 1933. The chemist was sure that the Germans would soon wake up and throw the bastards out.

Of course, another of the larger issues at play here—indeed, the central one in this book—is the West's, especially America's, response to the human tragedy unfolding in Nazi Germany and in Nazi-occupied Europe. The Hitler regime's racist policies caused occasional criticism from Western leaders and in the Western press, but there were no concerted efforts by the democratic governments to pressure the Nazis to alter their policies. In essence, the Jewish persecution was regarded as an "internal matter," off-limits to outside "interference."

Initially, the majority of German Jews who elected to escape the Nazi persecution through emigration sought sanctuary in other European countries or in Palestine, but as the oppression intensified a growing number chose to emigrate to the United States—or, more accurately put, they attempted to emigrate to the United States. In the end, the number of Jewish refugees who found safety in America was far less than the number of those who sought sanctuary there. Only about one-quarter of the Jews seeking to emigrate to America from the German Reich between 1933 and 1941 obtained visas under the quota system for Germany, which allocated 27,370 places annually. From Hitler's seizure of power until the closure of U.S. consulates in Germany and Austria in 1941, American consuls granted roughly 60,000 visas directly from the Nazi Reich. Over the thirteen-year period between 1933 and 1945 only about 35.8 percent of the German-Austrian quota was actually used, the quota being filled for the first time in 1939. Peak immigration occurred in the crucial eighteen-month period between March 1938 and September 1939, when some 45,210 ethnic German aliens, most of them Jews, entered the United States under the newly combined German-Austrian quota. Anti-immigration activists complained that the alien newcomers threatened to "flood" America, but the number was more like a trickle in a nation of some 130 million people. The far more significant number was the tens of thousands of Jews—from Germany and elsewhere in Nazi-occupied Europe—who were unable to join even the small stream of refugees flowing through the narrow sluice-gates of American immigration law during this time of crisis.

Severe limitation of immigration was a new phenomenon in the history of Jewish migration to the United States. Despite periodic upsurges of anti-Semitic sentiment in America during the late nineteenth and early twentieth centuries, German Jews, along with Eastern European Jews fleeing political oppression and economic misery in their homelands, flocked to the United States in large numbers. More than 2.5 million Jews found refuge in America during this period. Restrictive new immigration laws passed in the early 1920s, which established the national quota system, significantly slowed the tide of Jewish immigration, especially from Eastern Europe, but even in the 1920s roughly twice as many Jews gained entrance to the United States as in the decade between 1933 and 1943, when the average (from all sources) was only 15,284 a year. In the decade immediately preceding World War I, *eight times* as many Jews entered the United States as in the Nazi era. In other words, the prospects for gaining entrance to America were the slimmest at precisely the moment when that sanctuary was most desperately needed.

Much has been written about the Western democracies' policies toward Jewish immigration in the 1930s and during the Holocaust. Although most commentators have been critical of the West's unwillingness to give sanctuary to a larger percentage of Germany's—and later Europe's—Jews, there is no consensus in the scholarly literature regarding the number of Jews who might have been saved, the timeframe in which any rescue might have been possible, the degree to which Jewish organizations in the West might have been able to push political leaders toward a more pro-refugee position, and the motivation behind the restrictions on Jewish immigration. That being said, there has been a tendency among professional historians in recent years to move away from popular indictments of policy makers whose alleged "abandonment of

the Jews" made them "complicit" in the crimes of the Nazis, in favor of
an emphasis on the various constraints under which the policy makers
had to work.

Focusing as it does on one specific case, this book cannot provide definitive answers to all the big questions regarding the overall shape of
Western immigration policy during the Holocaust. As will become evident in the following discussion, however, my view is that one cannot
understand what happened to Max Schohl without being aware *both* of
a narrow-visioned (and sometimes anti-Semitic) policy-making apparatus in the Western democracies *and* a broader political-economic environment that militated against a substantially more liberal stance. One
must always bear in mind that in a time of severe depression and
chronic unemployment there was very little popular support in any of
the Western nations for throwing open the doors to masses of newcomers, however desperate they might have been. Yet in the end, my central concern here is not so much to "expose" the shortcomings of
American and British policy toward Jewish refugees during the Hitler
era as to examine the consequences of high-level public policy at the
"low" level of individual personal experience. What I try to do in this
narrative is to attach a specific human face and voice to the otherwise
bloodless record of political calculations and bureaucratic regulations.
The ordeal of the Schohls, and to a lesser extent that of their would-be
American rescuer, Julius Hess, can serve to remind us that abstract policies regarding national quotas, financial guarantees, and "ethnic balance" were anything but abstract to people whose own lives, and those
of their loved ones, were in peril.

While the broader parameters for this study include Nazi racial persecution, the dilemmas of assimilated German Jews, and the West's woefully inadequate response to the challenges of the Holocaust, most of the
action takes place on the tiny stage of a smallish German town—
Flörsheim-am-Main—during the crisis decades of the 1920s and 1930s.
It was in this modest town on the Main River near Frankfurt that Max
Schohl settled upon his return from World War I, and it was here that he
started his family and launched his career as a chemical entrepreneur.
Dr. Schohl quickly became a pillar of the community, someone his fellow

Flörsheimer could lean on in hard times. During the horrific inflation of the early 1920s he paid his workers in solid American dollars, and in the depression of the early 1930s he established a soup kitchen for the hungry. When Hitler took power in Berlin—and local Nazis accordingly seized control in Flörsheim—Max assumed that his past contributions to the community, along with his heroic war record, might afford him a certain protection. The reality was that he and the other Jewish businessmen in town were quickly reduced to pariah status: economically ruined, politically disenfranchised, socially marginalized, and subjected to constant harassment from officials and even ordinary townsfolk. During the infamous "Night of the Crystals" of November 1938, Max's house was thoroughly ransacked by Nazi thugs, some of whom had once been on the receiving end of his largesse. The Schohls' experience in Flörsheim, which essentially mirrored that of the other Jewish families in town, has much to tell us about the workings of Nazi politics at the grassroots level, about the evolution of the Holocaust from the bottom up.

A family history like this one is possible only if the author has access to a substantial body of firsthand information. The most important written record for this book is a unique collection of letters between members of the Schohl family (especially Max) and his American relatives (especially Julius Hess). The existence of correspondence from both sides of the Atlantic is a tremendous advantage. Whereas the letters from Max obviously give us a picture of what was happening in Germany, the letters from Julius Hess open up a window on the mental world of an "ordinary Joe" from the American provinces who was suddenly faced with a set of challenges unlike any he had ever known. Julius's sponsorship of Max's fruitless immigration suit brought home for him not only the horrors of Nazi racism but also the iniquities and capricious cruelties of America's policy toward Jewish refugees. Hess remained a patriotic American citizen throughout this ordeal (just as Max, more amazingly, remained a patriotic German to the bitter end), but he seems to have lost some of his unqualified faith in America's leaders and political institutions. Of course, in experiencing a negative epiphany as a result of prolonged contact with American officialdom Julius was taking part in yet another larger story, and one that is ongoing.

I would like to say that I "discovered" the cache of letters reproduced (generally in their entirety and always without grammatical corrections) throughout this book, but that is not the case. Some of the letters were excerpted in a *New York Times Magazine* piece by Michael Winerip entitled "Dear Cousin Max" (April 27, 1997). Winerip's article contained little commentary, and it occurred to me upon reading his piece that the letters he quoted could pack an even more powerful punch if placed within the framework of their various backstories, large and small. As I conducted research for this project I found other primary documents, including records relating to the Nazi era in Flörsheim, that give the Schohl family letters added meaning. To gain perspective on the American and British immigration policies that kept Max and thousands of other Jewish refugees at bay, I consulted, in addition to the wealth of secondary literature, various public and private document collections in U.S. and U.K. archives. For additional background information, and for many of the more intimate and personal details that help bring this story to life, I turned to witnesses of the Nazi era still living in Flörsheim, to local authorities on the history of that city, and above all to Käthe Schohl Wells, now the sole survivor among Max Schohl's immediate family.

Among the items that Käthe Schohl brought with her to America when she emigrated in 1946 was a framed inscription by Friedrich Schiller that now hangs in her kitchen in Charleston. It reads: "*Es gibt kein Übel so gross, wie die Angst davor*" ("No evil is as great as the fear of it") There is tremendous irony here, for it was, among other misjudgments, precisely an *inadequate* fear of the evil of Nazism that prompted Max Schohl to delay his efforts to get his family out of Germany. Yet at the same time one can understand why Käthe would prize this adage. Although Max was unable to deliver his wife and daughters from harm's way during the Nazi terror, his hope that they would be spared was ultimately realized, and that salvation was achieved in large part because neither he nor they became so overcome with fear that they gave up the fight to live. The fact that Max's family survived an evil that killed so many, and that his daughters were ultimately able to lead prosperous and fulfilled lives, lends the Schohl family story an element of redemption, even of transcendence.

consistent quality through hiring and training, 98–99
differentiation strategies based on service, 96–99
focus on interactions and relationships, 94–96
maintaining individual personality and creativity, 100–101
results of programmed behavior, 100
technology used to differentiate how staff serves guest, 150–153
technology used to shape employees' attitudes and behaviors, 153–155
Hotel Okura, 98
HSN. *See* QVC versus HSN
human capital, 18
human-machine interactions
cognitive attributes of a service interface, 125–129
emotional attributes of a service interface, 129–135
flowchart of interactions, 120f
integrating the attributes, 136–139
physical attributes as a service interface, 121–125
synaptic attributes of a service interface, 135–136
Hyatt Hotels, 117
hybrid interface systems. *See* people combined with machines

IBM Global Services, 36
Idei, Nobuyuki, 130
Immelt, Jeffrey, 36
industrial revolution, 2–3, 7–8
ING Direct, 77–78, 169
In-N-Out Burger, 87–88
interactions defined, 5, 6f
intelligence and interactivity, 40–43
interface, xvii–xviii, 5. *See also* Charles Schwab, First Direct, QVC
archetypes, 68f
defined, 249
interface advantage, 30, 36–37, 176
interface audit
evaluation questions, 212–215

purpose of assessment, 208–209
system inventory steps, 209–211
system scorecard (*see* scorecard for interface system)
interface imperative, 32
interface system, xviii, 251. *See also* scorecard for interface system
adaptability of 215, 216f, 225–226
defined, 251
interface technology
assessment approach (*see* interface audit)
capability as the basis of competitive advantage, 37
collaboration between humans and machines example, 29
commoditization and (*see* commoditization)
hybrid interfaces (*see* people combined with machines)
interface imperative, 52–53
machines' legacy of substitution for humans, 28–29
managing systems (*see* managing interface systems)
margin compression, 35–36
product life-cycle acceleration, 33–34
services as basis of revenues, 36
summary, 52–53
superior interface systems examples (*see* First Direct; QVC)
International Longshore and Warehouse Union, 56
Isen, Alice, 84
IT investment payback question, 17–19

Japan, 39
JetBlue Airways, 101, 102, 104
Jorgenson, Dale, 17

Kasparov, Garry, 118
Kawasaki, 41
Kismet, 45

Kramnik, Vladimir, 118–119
Krispy Kreme, 13
Kroger, 61
Kurzweil, Ray, 118

labor
  compensation index, 12
  outsourcing/off-shoring jobs, 21,
    51, 57, 58
  scarcity, 5–8 (*see also*
    outsourcing/off-shoring jobs)
  strikes, 56–57
Lands' End, 109
Levine, Kathy, 183–185
LG Home Shopping, 178–179
Luddites, 55

machine-dominant interfaces
  business implications of interactive
    machines, 140
  compensating for employee's lack of
    expertise or knowledge, 113–114
  described, 68–69
  dynamics of human-machine
    interactions (*see* human-machine
    interactions)
  economic implications of growth
    in machine-based jobs, 141–142
  future trends, 139–140
  human versus machine
    intelligence, 118–119
  potential of white-collar machines,
    117–118
  products as the faces and voices of
    brands, 114–115, 124
  role of emotion in interactions,
    140–142
  scope of tasks now done by
    machines, 115–117
  summary, 142–144
machines
  chess playing and, 118
  collaboration with humans
    example, 29
  company examples of front-office
    machines use, 13–14

contribution to humanizing the
  face of business, 15–16
contribution to productivity and
  customer base at First Direct,
  169–170
as the dominant interface (*see*
  machine-dominant interfaces)
emotional attributes and (*see*
  emotional attributes and
  machines)
in gaming industry (*see* gaming
  industry)
human response to a positive
  interface, 84
hybrid interfaces (*see* people
  combined with machines)
legacy of substitution for humans,
  28–29
machine-to-machine
  communication, 135–136
as service providers, 83–84
socially perceptive reasoning by, 45
task replacement of humans
  examples, 92
trained to act like people, 22
used to manage customer
  relationships, 9
machines supported by people
  human personality amplified by
    machines, 161–162
  leverage model, 159–161
  productivity model example (*see*
    First Direct)
  sources of value created, 158–159
managing interface systems
  affective appeal, 43–47
  device proliferation, 37–39, 40f
  intelligence and interactivity,
    40–43
  managers' neglect of systems as
    sources of competitive
    advantage, 176
  stages of system design (*see* Five A's
    model)
  summary, 204–205
  superior interface systems examples
    (*see* First Direct; QVC)
  supermarket systems, 173–174

system assessment (*see* interface audit)

tailoring systems to customer segments, 174–175

technology trends, 37

ubiquitous connectivity (*see* connectivity)

Mann, Steve, 146–147

margin compression, 35–36

McDonald's, 60, 155–158

Microsoft, 191

Midland plc. *See* First Direct

Minsky, Marvin, 130

MIT. *See* Affective Tigger; Brynjolfsson, Erik; Kismet, Eliza; Mann, Steve

Mitsubishi Heavy Industries, 41

Monte Carlo casino, 73, 76

Moore's Law, 37–38

Moravec, Hans, 118, 119, 130, 140

Narayanan, Shrikanth, 45

Nass, Clifford, 133, 134

National Car Rental, 116

Neeleman, David, 104

NeverLost, 116

Nordstrom, 94–96

Norman, Don, 84

Northwest Airlines, 14

off-shoring jobs. *See* outsourcing/ off-shoring jobs

online book sellers, 126–127

online commerce revolution, 126

online gaming, 48–49

OnStar System by GM, 51

Organisation for Economic Cooperation and Development (OECD), 3

organizational capital, 18

outsourcing/off-shoring jobs, 21, 51, 57, 58

overcapacity, 34–35

Palm V, 47

people combined with machines

human versus machine abilities, 146f

hybrid interface archetypes, 146

hybrid interface examples, 146–149

machines supported by people (*see* machines supported by people)

people enabled by machines (*see* people enabled by machines)

summary, 171–172

people-dominant interfaces

attributes, 89–91, 107–108

described, 67–68

difficulties managing meaningful interactions, 108–110

fast-food industry example, 87–88

localized connectivity in small businesses, 92–94

parallelism between human resource requirements and technology attributes, 91

relationship drivers, 90f

service on a limited scale (*see* hospitality industry)

summary, 110–112

task replacement by machines examples, 92

people enabled by machines

in fast-food industry, 155–158

in hospitality industry (*see* hospitality industry)

personalization paradox, 157

retail industry example, 149–150

personalization paradox, 157

personal video recorders (PVRs), 131

pervasive computing, 135–136

pharmacies, 61, 62

Picard, Rosalind, 140–142

Poland Spring, 129

Postrel, Virginia, 84

Predator Missile, 148–149

presentation layer and the interface system, 83–84

Procter & Gamble, 35

productivity gains using machines effectiveness driven by technology examples, 60–62

productivity gains using machines
  (continued)
  elements of productivity, 59
  IT investment and, 17–19
  parallels between manufacturing
    and services, 62–64
product life-cycle acceleration,
  33–34
Progressive Insurance, 14
proliferation. *See* device proliferation
push and pull technology, 121–124
PVRs (personal video recorders),
  131. *See also* TiVo

QRIO, 45–46
Qualia, 130
QVC
  backyard fence, 187–188
  basis of distinctiveness, 181–182
  brand-building successes, 192–193
  company background, 178–179
  customer base, 182
  customer satisfaction levels, 181
  engendering of trust, 189
  Five A's model applied to (*see* QVC
    versus HSN)
  focus on sales productivity per
    minute, 182–183
  interdependence of interface
    systems design, 190
  interface system flow, 211f
  interface system management in
    real time, 186
  leveraging of human talent,
    183–184
  leveraging of viewer loyalty
    through relationships, 191–192
  manipulation of productivity of the
    interface system, 184–185
  market value, 193–194
  order fulfillment record, 188–189
  philosophy of host interaction,
    187–188
  pricing strategy, 186–187
  product focus, 180–181
  selling platform, 179–180

QVC versus HSN
  activation and evolution of inter-
    faces and interface systems,
    203–204
  alignment of front-office
    capabilities, 201–202
  articulation of a plan of execution,
    202–203
  aspirations to identify best
    configuration of interfaces and
    interface systems, 199–201
  assessment of customers'
    interactions experiences,
    195–199
  order fulfillment record, 198
  pain points, choke points, drop-off
    points, 198–199
  purchasing process, 197
  in reach and resources, 194–195
  systems' interfaces, 195–197

Raskin, Jeff, 129
Recreational Equipment, Inc. (REI),
  60–61
Reed, John, 70
reengineering, 64–66
Reeves, Byron, 133, 134
REI (Recreational Equipment, Inc.),
  60–61
relationships defined, 5, 6f
retail industry, 149–150
Rifkin, Jeremy, 141
Rite Aid, 61
Ritz-Carlton Hotels, 99–100,
  150–153
Roach, Stephen, 17
robotics, 9, 12, 13f, 40–43, 131,
  137
Roomba, 41, 137
Ryanair, 101, 102

Sammons, Mary, 62
Sanyo Electric Co., 41
Schrobbie, 42
Scorecard at Fairfield Inn, 153–154

scorecard for interface system. *See also* adaptability of interface system; consistency of interface system; effectiveness of interface system; efficiency of interface system

adaptability evaluation, 225–226

adaptability metrics, 226f

characteristics of optimally configured interface system, 215–216

composite score analysis, 226–228

consistency evaluation, 222–225

consistency metrics, 224f

effectiveness evaluation, 217–218

effectiveness metrics, 219f

efficiency evaluation, 218–222

efficiency metrics, 220t, 221t, 223f

evaluation criteria, 216–217

factors influencing interactions, 218

Sears, 108–109, 193–194

Segel, Joseph, 179

self-checkout lanes, 173–174

service management challenges

hurdle rate for loyalty behavior change, 81–82

importance of quality of interactions with a company, 82

industrial versus service businesses, 78–79

requirement to satisfy efficiency and effectiveness, 82

service profit chain, 79

tradeoffs leading to diminished quality of interaction, 80–81

service profit chain, 79

service robots, 4, 9, 131, 137

service sector

attributes of a service interface (*see* human-machine interactions)

breakdown of costs to handle a customer inquiry, 11–12

differentiation strategies based on service, 96–99

economics of (*see* economics and the service sector)

machines as service providers, 83–84

management challenges (*see* service management challenges)

parallels between manufacturing and services, 62–64

people-dominant interfaces (*see* hospitality industry)

productivity gains from technology, 16–17

profit chain, 79

quality's influence on purchase decisions, 134–135

service at scale (*see* airline industry, low-cost)

service on a limited scale (*see* hospitality industry)

service recovery concept, 152

services as basis of revenues, 36

services' role in competitive advantage, 2

wealth creation through services, 3

Sharp, Isadore, 96–97, 98

Shopping Buddy, 174

Sinclair Broadcast Group, 160–161

slot machines, 73–74

small businesses and connectivity, 92–94

Solow, Robert, 17

Song by Delta, 107

Sony, 41, 130

Southwest Airlines, 29, 101, 102, 103–105, 106

Stealth Bomber, 148–149

strikes, 55–56

supermarket systems, 173–174

Sutter Health, 14

Taft-Harley Act, 56

Tellme, 132

three-six-one world, 31–32, 34

TiVo, 51, 131, 137

Toyota, 130

union protests against labor substitutions, 56

US Airways, 105

value equation, 59
value profit chain. *See* service profit
     chain
voice-recognition software, 129. *See*
     also Poland Spring; Tellme;
     Wildfire Communications

Wagamama, 147
Wakamaru, 41
Wal-Mart, 149–150, 193–194

Warner, Bill, 137
wealth creation through services, 3
Weather Services International
     (WSI), 159–160
Weizenbaum, Joseph, 133
Wildfire Communications, 137–139
World Robotics 2003, 12

Zara, 34
Zona, Joe, 159–161

About the Authors

Jeffrey F. Rayport is a former Harvard Business School Professor and is currently Chairman of Marketspace LLC, a strategic advisory, executive education, and software development business of Monitor Group, which is an international strategic advisory and investment firm headquartered in Cambridge, Massachusetts. At HBS for nearly a decade, Rayport developed and taught the first e-commerce course in the nation, authoring more than a hundred HBS case studies. Business plans produced by students in his course contributed to establishing many high-tech start-ups, including Yahoo! Prior to his leave from HBS, Rayport originated the concept of "viral marketing." He also became the only faculty member ever voted "outstanding professor" for three years running by the HBS Students Association.

Bernard J. Jaworski is Vice-Chairman of Marketspace LLC and President of its Monitor Executive Development business unit. He was previously a tenured member of the marketing faculty at the University of Southern California's Marshall School of Business and the Jeanne and David Tappan Marketing Fellow. From 1994 to 1999, he served as Dean of the Texas Instruments Virtual University. He is the recipient of both teaching and research awards, including the students' M.B.A. teacher of the year award at USC, the Alpha Kappa Psi award (twice) for best marketing-practice article in the *Journal of Marketing*, and the Jagdish Sheth Award (with Ajay Kohli) for long-term contribution to marketing theory and practice from the *Journal of Marketing*.

As partners in Monitor Group, Rayport and Jaworski have coauthored several market-leading business school textbooks focused on strategy and marketing in a networked economy, including *eCommerce* (2000), *Cases in eCommerce* (2001), and *Introduction to eCommerce* (second edition, 2003), which have been adopted at more than four hundred business programs around the world.

To
Käthe Schohl Wells
and in memory of
Dr. Max Schohl

—⚮—

Shall we refuse the unhappy fugitives from distress that hospitality which the savages of the wilderness extended to our forefathers arriving in this land? Shall oppressed humanity find no asylum on this globe?

THOMAS JEFFERSON, 1801

Hitler was quicker than the consuls on whose moods depended the visas that could save us.

ALFRED POLGAR

And you still thought, after the Nuremberg Laws and other horrors, that you were Germans? But we *were* Germans; the gangsters who had taken control of the country were not Germany—*we* were.

PETER GAY

Contents

Acknowledgments   xi

Introduction   xv

1   Max   1

2   "No Entry for Jews"   25

3   Paper Walls   53

4   "The Night of the Crystals"   91

5   High Hopes and Hot Tears   117

6   The Last Trial   167

Epilogue   217

Notes   235

Bibliography   257

Index   265

Acknowledgments

THE BOILERPLATE FORMULA in which an author swears that he could not have written his book without the help of many others is actually true in this case; this book could not even have been conceived, much less written, without the inspiration, guidance, and support of a host of individuals and institutions. Jody Hotchkiss, of Hotchkiss and Associates, New York, convinced me that a magazine piece published by Michael Winerip in the *New York Times* in 1987 constituted the raw material for a full-length historical study; my thanks to Jody for his prodding, and to Mr. Winerip for his original spadework. To translate the initial conception into a viable enterprise I needed the assistance of Dr. P. J. Wells, Max Schohl's grandson, who gave me access to the trove of family letters and documents that his mother, Käthe Schohl Wells, entrusted to him. P. J.'s wife, Fran, organized and catalogued the documents in a way that made them much easier to use. My agent, Agnes Krup, pushed the book along with her usual blend of patience and tenacity. I've had editorial help at Basic Books from Don Fehr and, after he moved on, from Sarah McNally, who made useful suggestions for the final draft.

For contacts, leads, and the literal opening of archive doors during my research in Flörsheim I am indebted to Bernd Blisch, formerly the head of that town's cultural office. Two other local experts on Flörsheim, Werner Schiele and Peter Becker, shared with me their immense knowledge of their town's history and traditions; moreover, their published works constitute an invaluable source for this project. For her memories of life in Flörsheim in the 1920s and 1930s I thank Frau Irmgard Radczuk, who was a neighbor of the Schohls in that era. Thanks also to Jan Radczuk for his own stories, as well as for the several bottles of vodka that fueled our conversations (and on whose influence any errors in my recollection of those stories can be blamed). Mr. Janun Wloral, who works as a guide at the Auschwitz concentration camp, kindly sent me documentation from the camp archive on the death of Max Schohl.

Work in many archives and libraries in America and Europe was indispensable for this book. I wish to acknowledge the staffs at the American Jewish Archives, Cincinnati, Ohio; the Hessisches Staatsarchiv, Wiesbaden; the Library of Congress Manuscripts Division, Washington, D.C.; the National Archives, Washington, D.C.; the Public Records Office, London; the Stadtarchiv, Flörsheim; the Wiener Library, London; the Yivo Institute for Jewish Research, New York; the archive of *The Charleston Newspapers*, Charleston, West Virginia (especially Mr. Bob Schwarz); the Landesbibliothek, Wiesbaden; the Leo Baeck Institute, New York; the University of California Library, Berkeley; the Green Library, Stanford University; the Hoover Institution, Stanford; the New York Public Library; and the Renne Library, Montana State University, Bozeman.

Professor Robert Rydell of Montana State University vetted the manuscript from his vantage point as an expert on modern American social and cultural history; my wife, Dr. Margaret Wheeler, vetted the manuscript from her vantage point as an expert on just about everything. My thanks to them both.

Another convention in acknowledgment writing is to save the greatest debt for last, and in this case the convention again has genuine meaning. Käthe Schohl Wells has been my partner in this project from beginning to end, sharing with me sometimes painful memories of her family's experiences during the early phases of the Holocaust in Flörsheim, of exile in Yugoslavia, and of slave labor in the war-torn Reich. She also painted vivid verbal portraits of the American relatives who tried to help her family escape to the United States during the Hitler era, and with whom she became personally close following her emigration to America after the war. To the degree that I have been able to bring the Schohl family and their would-be American rescuers to life, and thereby to endow this story with a vital human dimension, I have done so only with the constant assistance of my partner. This is her book as much as mine.

Introduction

"THE LAST TIME I saw my father," recalls Käthe Schohl Wells, "we were sitting on my bed in our flat in Ruma [Yugoslavia]. He said he would be back in a few days, and not to worry. We, my mother and sister and I, believed him, because he had been arrested several times before and always come back. We had no idea when they took him away that summer in 1942 that this time it was for good." A year later the Schohl women learned that the head of their family, Dr. Max Schohl, had been deported to Auschwitz, where he died in December 1943.

Käthe Schohl Wells is now a handsome and vital seventy-nine-year-old widow who uses her cane less as a walking aid than as a pointer with which to give directions. Although she has resided in the United States since emigrating from Germany in 1946 and is a proud American citizen, she comes across as a quintessential German-Jewish grandmother, which in fact she is. A visitor to her home is invariably treated to generous helpings of matzoth-ball soup followed by cake and coffee "mit Schlag." If Käthe takes a shine to her visitor, he'll likely come away with provisions for a week and a month's supply of Nürnberger Lebkuchen. Similarly, Käthe's comfortable apartment overlooking the Kanawha River in Charleston, West Virginia, is in many ways a little outpost of the country she left behind over a half-century ago. The china cabinet is filled with Meissen porcelain, and the walls are covered with pictures of her German family and the village near Frankfurt where she grew up, Flörsheim-am-Main. The most arresting photo in her picture gallery features a handsome young man in a German army officer's uniform from World War I; it is Käthe's father, Max, of whom she still has trouble speaking without her voice cracking. She says today that it is the ever-vivid memory of her father that inspires her to travel up and down the eastern seaboard giving talks to Jewish groups about her experiences in the Holocaust.

At the time of Max Schohl's final arrest the Schohl family had been living in exile in Yugoslavia for a little more than two years. They had fled there in March 1940 because as a Jew Max was no longer permitted to live and work in his own country, which he had served loyally and with distinction in World War I. Five years earlier he had lost his chemical factory due to the racist economic policies of the Nazis. In spring 1940 the Gestapo had given him a deadline to leave Germany immediately or face incarceration in a concentration camp. (In Nazi Germany, a policy of forced emigration preceded the program of mass murder in the evolution of the Holocaust.)

Although Yugoslavia provided a temporary haven for the Schohls, it was by no means their first choice as a refuge. Like many German Jews, Max and his family had hoped above all to emigrate to America, where Max believed that he could find work as a chemist, a field in which he had already distinguished himself in Germany. Moreover, Max had some relatives in America, including a cousin named Julius Hess, a clothing store clerk and part-time insurance salesman based in Charleston, West Virginia. Julius promised to help the Schohls obtain immigration visas to America. Max penned his first SOS appeal to "Cousin Julius" in September 1938, when the Nazi persecution of the Jews was already well in progress, and when prospects for finding refuge in America were becoming slimmer because more and more German Jews were applying for visas under a national quota system that remained highly restrictive. Nonetheless, both Max and Julius were confident that the Schohls' visa applications would be successful. After all, Max was a talented chemist and proven entrepreneur; his wife Liesel and his two daughters, eighteen-year-old Helene and fifteen-year-old Käthe, were all willing and anxious to work hard in their prospective new home. Were not Max and his family just the kind of new citizens that America needed?

After roughly two years of effort—of increasingly desperate attempts to find a way through or over the "paper walls" of American immigration policy—Max had to accept the grim reality that America did not, in fact, really need or want the Schohls. In 1939 the American consulate at Stuttgart allotted Max and his family a number within the annual German immigration quota (part of a racially based national quota system for all immigrants to the United States) that would not have allowed them to leave Germany for America until 1944, a delay that Max understandably believed would be fatal.

Despairing of reaching America, Max next sought to emigrate to England, where he had business contacts who promised to help him gain admission to that nation. But here too he ran into government-mandated impediments, which became insurmountable with the outbreak of war in September 1939. Facing the Gestapo-imposed deadline to leave Germany or face arrest, the Schohl family chose Yugoslavia as a temporary refuge in hopes of using that Balkan state as a springboard to emigration to Brazil. But before they could secure entry permits, Brazil too closed its doors. With the Nazi conquest of the Balkans in spring 1941, the Schohls' Yugoslavian refuge turned into a trap. A year after Max's deportation his wife and daughters were ordered back to Germany to work as slave laborers. They survived this ordeal, and by emigrating to America shortly after the war Käthe Schohl was finally able to make the passage that was denied to her father.

As the foregoing outline makes evident, the Schohl story constitutes a small part of a much larger story, or collection of stories. One of the bigger pictures here clearly involves the reaction of deeply assimilated German Jews to the gradually escalating Nazi racial persecution in their country. In putting off his bid to abandon Germany to such a late date, Max was allowing his love of country to obscure his survival instincts, a miscalculation that he shared with thousands of his co-religionists. Like all too many of them, he believed that the German people were too sophisticated, too cultured, to tolerate for long the barbarous brownshirted gangsters who had taken over his country in 1933. The chemist was sure that the Germans would soon wake up and throw the bastards out.

Of course, another of the larger issues at play here—indeed, the central one in this book—is the West's, especially America's, response to the human tragedy unfolding in Nazi Germany and in Nazi-occupied Europe. The Hitler regime's racist policies caused occasional criticism from Western leaders and in the Western press, but there were no concerted efforts by the democratic governments to pressure the Nazis to alter their policies. In essence, the Jewish persecution was regarded as an "internal matter," off-limits to outside "interference."

Initially, the majority of German Jews who elected to escape the Nazi persecution through emigration sought sanctuary in other European countries or in Palestine, but as the oppression intensified a growing number chose to emigrate to the United States—or, more accurately put, they attempted to emigrate to the United States. In the end, the number of Jewish refugees who found safety in America was far less than the number of those who sought sanctuary there. Only about one-quarter of the Jews seeking to emigrate to America from the German Reich between 1933 and 1941 obtained visas under the quota system for Germany, which allocated 27,370 places annually. From Hitler's seizure of power until the closure of U.S. consulates in Germany and Austria in 1941, American consuls granted roughly 60,000 visas directly from the Nazi Reich. Over the thirteen-year period between 1933 and 1945 only about 35.8 percent of the German-Austrian quota was actually used, the quota being filled for the first time in 1939. Peak immigration occurred in the crucial eighteen-month period between March 1938 and September 1939, when some 45,210 ethnic German aliens, most of them Jews, entered the United States under the newly combined German-Austrian quota. Anti-immigration activists complained that the alien newcomers threatened to "flood" America, but the number was more like a trickle in a nation of some 130 million people. The far more significant number was the tens of thousands of Jews—from Germany and elsewhere in Nazi-occupied Europe—who were unable to join even the small stream of refugees flowing through the narrow sluice-gates of American immigration law during this time of crisis.

Severe limitation of immigration was a new phenomenon in the history of Jewish migration to the United States. Despite periodic upsurges of anti-Semitic sentiment in America during the late nineteenth and early twentieth centuries, German Jews, along with Eastern European Jews fleeing political oppression and economic misery in their homelands, flocked to the United States in large numbers. More than 2.5 million Jews found refuge in America during this period. Restrictive new immigration laws passed in the early 1920s, which established the national quota system, significantly slowed the tide of Jewish immigration, especially from Eastern Europe, but even in the 1920s roughly twice as many Jews gained entrance to the United States as in the decade between 1933 and 1943, when the average (from all sources) was only 15,284 a year. In the decade immediately preceding World War I, *eight times* as many Jews entered the United States as in the Nazi era. In other words, the prospects for gaining entrance to America were the slimmest at precisely the moment when that sanctuary was most desperately needed.

Much has been written about the Western democracies' policies toward Jewish immigration in the 1930s and during the Holocaust. Although most commentators have been critical of the West's unwillingness to give sanctuary to a larger percentage of Germany's—and later Europe's—Jews, there is no consensus in the scholarly literature regarding the number of Jews who might have been saved, the timeframe in which any rescue might have been possible, the degree to which Jewish organizations in the West might have been able to push political leaders toward a more pro-refugee position, and the motivation behind the restrictions on Jewish immigration. That being said, there has been a tendency among professional historians in recent years to move away from popular indictments of policy makers whose alleged "abandonment of the Jews" made them "complicit" in the crimes of the Nazis, in favor of an emphasis on the various constraints under which the policy makers had to work.

Focusing as it does on one specific case, this book cannot provide definitive answers to all the big questions regarding the overall shape of Western immigration policy during the Holocaust. As will become evident in the following discussion, however, my view is that one cannot understand what happened to Max Schohl without being aware *both* of a narrow-visioned (and sometimes anti-Semitic) policy-making apparatus in the Western democracies *and* a broader political-economic environment that militated against a substantially more liberal stance. One must always bear in mind that in a time of severe depression and chronic unemployment there was very little popular support in any of the Western nations for throwing open the doors to masses of newcomers, however desperate they might have been. Yet in the end, my central concern here is not so much to "expose" the shortcomings of American and British policy toward Jewish refugees during the Hitler era as to examine the consequences of high-level public policy at the "low" level of individual personal experience. What I try to do in this narrative is to attach a specific human face and voice to the otherwise bloodless record of political calculations and bureaucratic regulations. The ordeal of the Schohls, and to a lesser extent that of their would-be American rescuer, Julius Hess, can serve to remind us that abstract policies regarding national quotas, financial guarantees, and "ethnic balance" were anything but abstract to people whose own lives, and those of their loved ones, were in peril.

While the broader parameters for this study include Nazi racial persecution, the dilemmas of assimilated German Jews, and the West's woefully inadequate response to the challenges of the Holocaust, most of the action takes place on the tiny stage of a smallish German town—Flörsheim-am-Main—during the crisis decades of the 1920s and 1930s. It was in this modest town on the Main River near Frankfurt that Max Schohl settled upon his return from World War I, and it was here that he started his family and launched his career as a chemical entrepreneur. Dr. Schohl quickly became a pillar of the community, someone his fellow

Flörsheimer could lean on in hard times. During the horrific inflation of the early 1920s he paid his workers in solid American dollars, and in the depression of the early 1930s he established a soup kitchen for the hungry. When Hitler took power in Berlin—and local Nazis accordingly seized control in Flörsheim—Max assumed that his past contributions to the community, along with his heroic war record, might afford him a certain protection. The reality was that he and the other Jewish businessmen in town were quickly reduced to pariah status: economically ruined, politically disenfranchised, socially marginalized, and subjected to constant harassment from officials and even ordinary townsfolk. During the infamous "Night of the Crystals" of November 1938, Max's house was thoroughly ransacked by Nazi thugs, some of whom had once been on the receiving end of his largesse. The Schohls' experience in Flörsheim, which essentially mirrored that of the other Jewish families in town, has much to tell us about the workings of Nazi politics at the grassroots level, about the evolution of the Holocaust from the bottom up.

A family history like this one is possible only if the author has access to a substantial body of firsthand information. The most important written record for this book is a unique collection of letters between members of the Schohl family (especially Max) and his American relatives (especially Julius Hess). The existence of correspondence from both sides of the Atlantic is a tremendous advantage. Whereas the letters from Max obviously give us a picture of what was happening in Germany, the letters from Julius Hess open up a window on the mental world of an "ordinary Joe" from the American provinces who was suddenly faced with a set of challenges unlike any he had ever known. Julius's sponsorship of Max's fruitless immigration suit brought home for him not only the horrors of Nazi racism but also the iniquities and capricious cruelties of America's policy toward Jewish refugees. Hess remained a patriotic American citizen throughout this ordeal (just as Max, more amazingly, remained a patriotic German to the bitter end), but he seems to have lost some of his unqualified faith in America's leaders and political institutions. Of course, in experiencing a negative epiphany as a result of prolonged contact with American officialdom Julius was taking part in yet another larger story, and one that is ongoing.

I would like to say that I "discovered" the cache of letters reproduced (generally in their entirety and always without grammatical corrections) throughout this book, but that is not the case. Some of the letters were excerpted in a *New York Times Magazine* piece by Michael Winerip entitled "Dear Cousin Max" (April 27, 1997). Winerip's article contained little commentary, and it occurred to me upon reading his piece that the letters he quoted could pack an even more powerful punch if placed within the framework of their various backstories, large and small. As I conducted research for this project I found other primary documents, including records relating to the Nazi era in Flörsheim, that give the Schohl family letters added meaning. To gain perspective on the American and British immigration policies that kept Max and thousands of other Jewish refugees at bay, I consulted, in addition to the wealth of secondary literature, various public and private document collections in U.S. and U.K. archives. For additional background information, and for many of the more intimate and personal details that help bring this story to life, I turned to witnesses of the Nazi era still living in Flörsheim, to local authorities on the history of that city, and above all to Käthe Schohl Wells, now the sole survivor among Max Schohl's immediate family.

Among the items that Käthe Schohl brought with her to America when she emigrated in 1946 was a framed inscription by Friedrich Schiller that now hangs in her kitchen in Charleston. It reads: "*Es gibt kein Übel so gross, wie die Angst davor*" ("No evil is as great as the fear of it") There is tremendous irony here, for it was, among other misjudgments, precisely an *inadequate* fear of the evil of Nazism that prompted Max Schohl to delay his efforts to get his family out of Germany. Yet at the same time one can understand why Käthe would prize this adage. Although Max was unable to deliver his wife and daughters from harm's way during the Nazi terror, his hope that they would be spared was ultimately realized, and that salvation was achieved in large part because neither he nor they became so overcome with fear that they gave up the fight to live. The fact that Max's family survived an evil that killed so many, and that his daughters were ultimately able to lead prosperous and fulfilled lives, lends the Schohl family story an element of redemption, even of transcendence.

16053